Like Thomas Friedman showe
how information is also bec
globalized. Friedman showed
two or more systems capable o.
ability of two systems to exchange information, O'Dell demonstrates the next stage in information sharability – Semantic interoperability. Semantic interoperability is the ability to interpret the information exchanged meaningfully and accurately in order to produce useful results.

Who will most benefit from data interoperability? O'Dell convincingly predicts that the biggest beneficiaries will be institutions and their constituents offering e-Government Services, e-Law Enforcement, and e-Health Care Services. *Silver Bullets* is recommended for those who seek not only to understand data interoperability, but also to capitalize on its predicted benefits.

—Neil R. Evans
Former CIO, Microsoft Corporation
Contributing Author, *Technology Everywhere – A Campus Agenda for Educating and Managing Workers in the Digital Age* (2002),
Educause Leadership Strategies, John Wiley & Sons, Inc.

Pete O'Dell's good book boils down to one main point: Standardized, interoperable data is the answer to a host of questions, and it can't come too soon. This is a book for the business or operations executive who wants to shake hands with the future rather than be run over by it. Pete shows how standards support civilization in every area from transportation to health care, and maps the way ahead with clarity, good will and humor.

—Bill Lang
Author of *Scores on the Board: The 5-Part System for Building Skills, Teams and Businesses*

Managing risk involves effectively managing data, finding its patterns while keeping information secure. *Silver Bullets* lays out a strong case for interoperable data systems that save time and money, and reduce risk.

—Annie Searle
Annie Searle & Associates Risk Consultants

This book is for all of us who live in the real world where information must be shared and acted upon, not just for the techno geeks who understand ones and zeros. As the Commander of U.S. Forces in Japan during the U.S. response to the horrific tsunami that took over 350,000 lives in the Indian Ocean on 26 December 2004, I was responsible for a bulk of the support that we provided to the senior U.S. military commander deployed to Thailand. We quickly found out that in spite of our ability to provide the President with streaming video of our operations, we could not communicate with NGOs, PVOs or even the Indonesian military, using our protected communications systems. As we tried to meet the humanitarian needs of those affected by the tsunami, we found that we could not communicate with those who knew what those needs were. We solved this problem with an inelegant solution: We established a Yahoo.com account that enabled us to match resources against humanitarian requirements. Had we been able to share interoperable data with those who had it, our task would have been more complete. Kudos to Pete O'Dell for issuing this wake-up call. Adoption of his vision could have saved lives then, and could save lives in the future.

—Lt. Gen. Thomas C. Waskow, Ret.
Commander, U.S. Forces Japan;
Commander, 5th Air Force, Yokota Air Base, Japan

Pete O'Dell builds a convincing case that all sectors of business activity thrive when they adopt standards to streamline commerce and internal operations. He illustrates how far such adoption has to go in the data field, then details how a variety of open-source standards are capable of dramatically speeding the use of interoperable data models globally, across sectors. When that happens, every aspect of business will be simplified and linked!

—W. David Stephenson
Stephenson Strategies; Author, *Democratizing Data* (coming Summer 2010)

One of the greatest challenge for marketers today is how to access, harness and leverage data to optimize revenue and earnings. It ain't easy; in fact it's bloody hard. Pete O'Dell gets it, and with Silver Bullets, he challenges CXOs to focus on data in a new way. Get excited about interoperable data; your cash flow goals may depend on it in our new digital world.

—Tim Carroll
Global Chief Marketing Officer, Village Roadshow, Melbourne, Australia

SILVER BULLETS

Also by Pete O'Dell:

The Computer Networking Book

SILVER BULLETS

How Interoperable Data
Will Revolutionize
Information Sharing
and Transparency

Pete O'Dell

authorHOUSE®

AuthorHouse™
1663 Liberty Drive
Bloomington, IN 47403
www.authorhouse.com
Phone: 1-800-839-8640

© *2010. Pete O'Dell.*
 Book design by Sherry Lamoreaux
 Title design by Justin Buckley
 All rights reserved.

No part of this book may be reproduced, stored in a retrieval system, or transmitted by any means without the written permission of the author.

First published by AuthorHouse 03/19/2010

ISBN: 978-1-4490-4075-8 (sc)
iSBN: 978-1-4490-4076-5 (hc)
ISBN: 978-1-4490-4077-2 (e)

Library of Congress Number: 2010903540

Author Photograph by Pam Angelus

Printed in the United States of America
Bloomington, Indiana

This book is printed on acid-free paper.

Dedication

To Pam Angelus for so many reasons. Thank you for all your love, support, and encouragement. This book literally would not have happened without you.

Special thanks

Special thanks to Sherry Lamoreaux for her tireless editing. I still don't know when a dash or an em-dash makes sense, but I trust they are in good order.

Another very special thank you to Jim, Lynn, Tim, and John. You were there at just the right moment, and I'm deeply appreciative.

*Form follows structure;
structure doesn't follow form.*

— I.M. Pei, architect

Contents

Foreword	xiii
Acknowledgements	xv

Chapter One
A major opportunity to improve civilization — 11

An Opinion
The top ten benefits resulting from the broad adoption of interoperable data — 31

Chapter Two
Long-ago standards: things you'll recognize — 37

Case Study
Jefferson Parish, Louisiana tackles Hurricane Gustav — 47

Chapter Three
Near-past lessons: standardized technology and computing — 49

Chapter Four
The beginning of standardized data transactions — 61

Case Study
Using interoperable data to improve diabetes management — 71

Chapter Five
The emergence of Extensible Markup Language; (XML); a tour of major initiatives — 73

Chapter Six
Two great interoperable data standards:
CAP and KML — 91

Case Study
Golden Phoenix — 109

Chapter Seven
Interoperable data efforts: things you can watch, use, leverage, and adopt — 115

Chapter Eight
Executing a data interoperability pilot (*right now*) — 131

Case Study
External alarm interface exchange standard — 141

Chapter Nine
The far-flung future: where do we go from here? — 145

Chapter Ten
Raves and conclusions — 163

Appendix A
Chapter notes and suggested reading — 175

Appendix B
Best of NIEM Awards 2009 Inaugural — 187

Appendix C
World Meteorological Organization Members — 193

About the Author — 195

Index — 197

Foreword

Sharable data can change the world. The web, at its root, is simply a standard protocol for data and virtually the whole network as we use it today has arisen from a similar source. Successful data standards come to be taken for granted, but building them takes work, foresight, and both technical and political leadership. Pete O'Dell combines a high-level analysis of the function of data standards with detailed, telling examples, and his book is both a guide and an inspiration to the people who are setting out to reform data sharing systems in government and business.

—Gary Wolf
Contributing Editor, *Wired* Magazine

Acknowledgements

I have the privilege of having very smart friends, associates, and volunteers who were kind enough to read this book while it was in progress. The book is better because people offered feedback and suggestions, debated points with me, argued about placement and readability, researched various facts, drew diagrams, argued national security issues, contrasted organizations large and small, and generally set me straight on a wide variety of topics.

I owe you all a gigantic thank you and a beer of your choosing! A few of these generous people are:

Lynn James; Neil Evans; David Stephenson (Stephenson Strategies); David Bray; Mike Helfrich; Laurent Liscia and Carol Geyer (OASIS); Eliot Christian (World Meteorological Organization); Tom Waskow; Yogesh Kelly; Brian Wilczynski (DoD); Laura White; Laura McNulty; Betty Levine (ISIS-Georgetown); Elysa Jones (OASIS, Warning Systems, Inc); Mano Marks and Phil Dixon (Google Federal); Chuck Gottschalk; John Ballantine; Mat Jennings; David Jones; Lyle Peters; George Arnold (DoE); Kshemendra Paul (OMB); Ann Castleton; Annie Searle; Len Hawkins; Don Harrison; Marc Clifton; James Dyche; Tim McCallum; Justin Buckley; Yael Schonfeld Abel; Annie Searle; and Sherry Lamoreaux.

Introduction

Bottom line up front (BLUF)

Interoperable data will revolutionize information systems and information sharing between organizations! You heard it here first.

We are in the middle of a quiet but important period of change, innovation, and standardization. Interoperable data will shift computerized systems from being software- and system-centric to being data-centric and customer-controllable – not to mention customer-created, with a spate of new Web 2.0 tools. The use of interoperable data will help enable transparency, accessibility, and usability of structured information of all types. Systems will evolve from application silos and stovepipes to interconnected, information-sharing data networks.

A quick definition of "interoperable data": As our basis for discussion, use the idea of an open, formatted set of data fields that define an action or a transaction; a good example is a credit card charge. Name, account information, date, merchant, and amount are all interoperable data that can be exchanged among vendors and banks. Another simple transaction is a tornado alert with the date, time, location, status, a map, and level of urgency. The alert can be created and published in an agreed-upon open format that is known to and consumable by many different systems and organizations.

Silver Bullets

The myth of the Silver Bullet

The concept of a "Silver Bullet" – one single strategy, technique or trick that can permanently resolve a persistent difficulty – began centuries ago as a charming bit of fiction. In recent decades, the phrase has become a buzzword for anything that promises that it can, by itself, easily slay some monster of a problem. In the information technology world, the Silver Bullet promise has failed so spectacularly that IT people cringe whenever the phrase is mentioned.

- From the fairy tales of the Brothers Grimm to today's popular fiction, literature tells us that a single silver bullet to the heart can permanently kill the most persistent vampire (or werewolf, witch or monster).

- Remember the Lone Ranger and Tonto? The Lone Ranger used a silver bullet to win the day, but still had to get out of a tricky situation or have Tonto save his butt at the last minute. Not a model for a modern-day technology project.

- In his timeless book, *The Mythical Man-Month: Essays on Software Engineering*, Frederick Brooks sets up the concept of the Silver Bullet as a game changer, then demolishes it. "There is no single development, in either technology or management technique which by itself promises even one order of magnitude improvement within a decade in productivity, in reliability, in simplicity." He's absolutely right. The single-solution approach has been the cause of major disasters in computing. Over software's short lifespan, a host of solutions have been held up as Silver Bullets: subroutines, structured programming, databases, object-oriented programming, and so on. Not one, by itself, has delivered.

Introduction

The difference (many) Silver Bullets can make. This book's purpose is to illustrate how a certain single solution – interoperable data – and way of thinking, if implemented across broad IT strata, can become a great strategic tool (if not a singular Silver Bullet) for an organization or industry.

The cumulative effect of many interoperable transactions (multiple Silver Bullets employed in multiple organizations) will be very powerful, and if done right, will result in a strategic leap forward over a relatively short time. Much like a single grain of sand, there is no initial impact when the first tiny bit of silicon is deposited, and a small collection can be washed away easily. But continue to add millions and billions more grains of sand, and in short order you've created the greatest beaches in the world. **It is the cumulative power of interoperable data transactions that we'll explore, and how that interoperability will change the world around you.**

Making use of interoperable data is an opportunity for every organization, big or small. Key benefits can include:

- Enabling information sharing with trusted partners
- Enhancing system capabilities and longevity
- Lowering overall costs of information applications
- Improving the breadth and quality of information
- Increasing the speed and accuracy of decisions
- Improving transparency and speed of disclosure of information to valid constituents
- Preserving data for future uses, including those not yet conceived

Silver Bullets

You can find countless examples of the sad state of data and information sharing as it's practiced (or not) today, and there is strong agreement on the need for change. The imperative for streamlining, speed, and innovation is increasingly important as the world tries to connect more and more data together and make sense of it.

Civilization has largely moved forward as a result of other interoperable and standardization efforts. Pioneers have been imagining and implementing various aspects of interoperable data for many years with good results. Like most revolutions, things develop almost unnoticed for a number of years. Then, when a tipping point occurs, the need for change becomes obvious and adoption grows very quickly. We're moving close to that tipping point with interoperable data.

> **An exception to the Silver-Bullets-Don't-Work rule: the Salk vaccine**
>
> Until 1955, polio was the most frightening public health problem of the postwar U.S. In 1952 there were over 300,000 cases and 58,000 deaths, mostly children. Researcher Jonas Salk created what can be hailed as a genuine Silver Bullet – a single vaccine that could prevent polio in most cases. Soon after Salk's vaccine was licensed in 1955 children's vaccination campaigns were launched. The annual number of polio cases in the U.S. fell to 5,600 by 1957 and to 161 in 1961.
>
> Salk refused to patent the vaccine, preferring to see it disseminated as widely as possible, as quickly as possible. When asked who owned it, Salk replied: "There is no patent. Could you patent the sun?"

Introduction

Key issues

Lifetime of data = forever. A key element in this interoperable data equation is time; your data will be around for a long time once it's in a system, likely outliving several successive different applications, or being repurposed in ways your organization has not thought of yet, in applications yet to be developed. Your data will outlive you – sorry. I'm sure the cavemen in France, had they known how many years their cave paintings would last, would have done a better job on the colors, and documented their work more.

If you or your organization have ever experienced the pain of transitioning from one computerized application package to another (perhaps having to reformat mountains of data to fit the new system), you'll understand why standardized data and planning ahead for the future are so important.

Vendor independence. If you choose standard formats, your organization can make software vendors come to you on a level playing field. Choose a proprietary format, and the vendor has you locked in (and locked out of working with other vendors or helping yourself).

Large numbers make things work. It is rumored that Alexander Graham Bell showed the first telephone to a senior executive of a telegraph company. After patiently explaining the technology, Bell asked the usual post-presentation question, "So, what do you think?"

The executive purportedly said, "So? Who am I going to call?" Millions of telephones later, the opportunity for the telephone – based on plurality – is apparent to all.

Who can benefit by reading this book?

Anyone involved with or impacted by information that's collected and managed by computerized systems (over one billion people around the world right now – with the next billion on the way).

If you're responsible for assimilating information from multiple sources; analyzing collected information; or distributing information to others inside or outside of your organization, this book may open new areas of thought and maybe make your life a little easier.

If you are one of the millions of people frustrated with the current state of your information systems and infrastructure, this book presents questions to ask and approaches to take. What you learn should give you hope that the state of the art will improve with the applications of standardization and interoperability.

If you want to push government or other large organizations to be more open and responsive, interoperable raw data is a key technique to enable this capability.

The job titles below represent people who could find tactical and strategic value in this information:

- Program managers
- Project managers
- Operations managers and executives
- Enterprise and information architects
- Business analysts and strategic consultants
- Business development executives
- Computer application managers and specialists
- Organizational managers across all functions and types – government, non-profit, commercial

Introduction

If you remember the time before the World Wide Web burst on the scene, you'll remember that getting information from someone else's computer system was difficult or very time consuming. "Sneaker-net" (walking around and moving floppy disks between disparate systems) was, regrettably, often the answer.

If you wanted to find out where your shipped package was located, you got on the phone with FedEx, not on a computer to look it up yourself. If you wanted a new book, you could drive down to your local book store or order it over the phone or through a paper catalog – no quick, one-click trip to Amazon.com.

Interoperable data will bring the same kind of "AHA!" moments that the World Wide Web did, when we all suddenly realized we could connect to people, communities and ecommerce sites, almost at will, in the early days of the browser and HTML. In a similar vein, remember what it was like trying to meet a friend at a concert before cell phones were pervasive?

When you start getting relevant, standardized, real-time information from around the world and combine it with your own internal data and that of your partners, you'll wonder how and why you were so blind for so long.

If you are a computer architect or developer, you'll find this book short on detailed technical and actual programming advice, but long on the important concepts of why interoperable data should be part of your architectural thinking and how interoperable data will enable you to connect to an even larger set of information sources and other systems that will make your organization more effective. It will give you a good basis for discussion and understanding of the issues that the operational and strategic people in your organizations are thinking about.

Many times, avoiding new system development and using standardized, off-the-shelf capabilities will speed new functions to production, lower costs, and give you an ongoing systems benefit from the creators and maintainers of these standardized approaches.

What this book isn't

There is no discussion of land-mobile radio interoperability in this book. Radio always presents a paradox to me. Making a voice-based "party line" able to handle hundreds or thousands of participants, all having to listen intently and wait for an increasingly small window in which to say something, has never seemed worth the billions spent by the various homeland security groups. It's clear that there is greater value to more people in the digital movement of structured data that can happen in near-real time, and be consumable and understandable by many organizations. "Say again, over" is not part of the interoperable data equation presented in this text.

There also isn't anything in the book about unstructured search – important as this is for today's web, my prediction is that structure and metadata, context, and semantics will eclipse unstructured data searching as it develops and becomes the mainstream method of storing information.

As important as the Semantic Web is for future generations, there is no attempt to explain it in detail, although interoperable data is one good step toward this smarter, more intelligent, next-generation paradigm concerned with structured data on the web.

Declared bias. The company I helped found after 9/11 – Swan Island Networks (www.swanisland.net) – produces a cloud-based, Web 2.0 managed service called TIES (Trusted Information Exchange

Service) which utilizes many of the principles described in this book. Swan Island learned many of the data interoperability lessons first hand and, as a pioneer in this space, has seen the enormous potential of aggregating disparate information sources together using interoperable formats. Many of the examples presented in this book are lessons gleaned from designing, developing, and deploying our situational awareness capability. One of the design tenets of TIES is interoperability and a "systems of systems" approach – not proprietary lock-in.

Second declared bias. Toward action! Making something happen! The number of good things bottled up by perpetual "analysis paralysis" is enormous, and there's a sad lack of even trial implementations. Innovation is hard, but it is one of the things that America has been famous for; we're starting to lag – especially in large organizations and government. Nothing takes the place of the actual implementation process, and the sooner you get underway, the better. Even if you fail at parts of it, starting the journey is important – the lessons learned are invaluable.

A note about technology, and "being technical"

It's a common misconception that to understand information processing you somehow need to be "technical" – a geek, programmer, developer, etc. This situation has been propagated largely by the people who are "technical" – it can be a convenient dodge for budgeting discussions, hard questions, project failures, and more. I've spent my life on both sides of the equation, first as a completely non-technical person - a restaurant manager, then a soldier in the U.S. Army Infantry. I went to school for programming and information technology under the GI Bill (thanks, America – this was a life changer!), then worked for years as a programmer, systems analyst, project leader, and finally, Chief Information Officer for several companies.

I came to realize that while I liked the technology, the real action (organizational impact, new product development, profit making) was all on the front lines of business that the IT groups supported. I've spent the rest of my career as a business guy who understands technology, or, at least the principles behind it, and how to ask the right questions. Don't let the technology snow you – it can be simplified and explained. Insist upon clarity and keep looking until you find someone who can translate, if necessary. If you are making organizational decisions with the *why* of technology in mind, you need to know the basic principles of *how*.

Chapter One

A major opportunity to improve civilization

We're only at the beginning of what we have to do here.
—Bill Gates

We live in an accelerating world of information, innovation, and networks, with a rapidly increasing number of "Netizens" – people who communicate and interact with others through the Internet – for business, government, personal, and other purposes. More people can communicate directly with each other in real time today than in the cumulative history of man. An event that 30 years ago would have captured local front-page newspaper headlines can whiz by almost unnoticed today, because so many other incredible things are happening around the globe in parallel. Newspapers themselves are losing relevance as people increasingly use the web to gather and sort information from a broader range of new sources. When something exciting happens, we want to learn about it *right now.* The Internet gives us that immediacy. Just as important, we can participate; we can exchange comments and share knowledge about what's happening around the world, in real time.

My life span is a typical illustration of this high-speed rate of change. I grew up on a dairy farm in upstate New York, where our main communication methods were a telephone party line (which we shared with six other families) and the U.S. mail. Local access could be challenging if there were young girls on the party line, but everyone typically had a way to signal others if access was important. Long distance calls were tremendously expensive relative to the incomes of the day, so only emergencies and special events triggered a phone call beyond the local calling area.

Compare and contrast that existence with today – a mere 45 years later. I have at least three phone numbers, multiple email addresses, Skype, text messaging, instant messaging, Twitter, a rarely used Facebook account, a heavily used LinkedIn account, and more – most with free, unlimited, unconstrained usage plans. That's just to reach me. The world that I can reach is virtually unlimited, and I'm the weakest link; I have to figure out what I need and allocate my consumption to fit within my ever-more-jammed 24 x 7 days.

My friends and business associates span the world, so email that's important can show up anytime, not just during normal business hours – if "normal" exists for anyone anymore. Mix in your social life, and you have a recipe for a complicated soufflé of capabilities and needs. Change the recipe slightly by subscribing to too many or irrelevant information feeds, and the whole information collection recipe turns into overload – it's a very delicate balance.

Even mere existence seems more complicated now. As I write this, we're in dangerous financial and economic times. Worldwide disasters are increasing, from deadly weather events (Hurricane Katrina, the 2010 Haitian earthquake) to man-made threats (the Mumbai terrorist attacks, the Madrid train bombings, the London Underground bombings, the Fort Hood shootings) to potential global health crises (the potential H1N1 pandemic)...with undoubtedly more by publication time.

Chapter One

A recent congressional report predicted that we'll have either a biological weapon attack or a nuclear attack by terrorists by 2013 if we don't take precautions to prevent such acts. Statistically, your odds of survival as an individual are very good, but few of us take comfort in that. It's much more human to be concerned about acute events – a sudden, unexpected threat from an improvised explosive device (IED), a sniper, or natural disaster – than to pay attention to the chronic problems to which we grow accustomed. To use health as an analogy, people often put off dealing with a chronic high-risk factor (e.g., high blood pressure that may led to a fatal heart attack) but respond immediately to an acute event (e.g., a broken leg in a car accident). Interoperable data can help mitigate some of the acute risks, but you'll still have to watch your diet and go for a run.

Humans are the weakest link. As this accelerating technology unfolds, we find that humans are still the weakest link in the information processing chain. Many of us are literally drowning in information on multiple fronts. This problem manifested when the Internet first came into the mainstream – the number of web sites, email addresses, ecommerce opportunities and web resources exploded, and finding anything was tough, even for early adopters.

Search engines were invented to help with the problem of finding what you want and ignoring the overload, and few of us go many days (or minutes) without using Google. (Google is the winner in the unstructured search engine wars, at least for today. One false step, or missed technological generation, and Google will be yesterday's news, with a new company in the limelight. If you remember Yahoo, Magellan, and several of the other search engines that once seemed to have an unbeatable position in this market, it's easy to see how fast a market leader can come and go.)

More and more information is also generated by machines, from sensors to transactional systems, automated feeds such as Really Simple Syndication (RSS), and others. Sensors alone will have an exponentially expanded presence in the coming years. Far lower costs, greater capabilities, and faster and more broadly available wireless networks for data collection will contribute to an explosion of sensor data that may or may not be publicly available, useful, or timely, but will be out there nonetheless. The potential gains that could be realized for harnessing sensor-generated information flow are huge – similar in scope to harnessing rivers to produce electricity through hydropower. Sensors almost uniformly use structured data formats for communication.

With better information, you can make enhanced decisions. With faster information, you can make accelerated decisions. Combine these two critical elements, and the improvement effect is multiplied versus additive. Reducing the elapsed time between an event's onset and your organization's decision about responding to it is usually very important, whether the event is foreseen or a complete surprise.

The major problem is getting the right information at the right time in the right format, on the right device, so that it can affect decision-making for the better when processed by the right human decision-maker. The solution (or the mission, if you prefer) is to use technology to improve and simplify emergency and business communications, while minimizing data overload and the opportunity for human error. Interoperable, structured data is a key to accomplishing this mission. How we move forward with standardizing data and its associated issues is a topic that will evolve with time and have different implications for people depending on their context and position, all impacted by a broad set of individual variables.

Chapter One

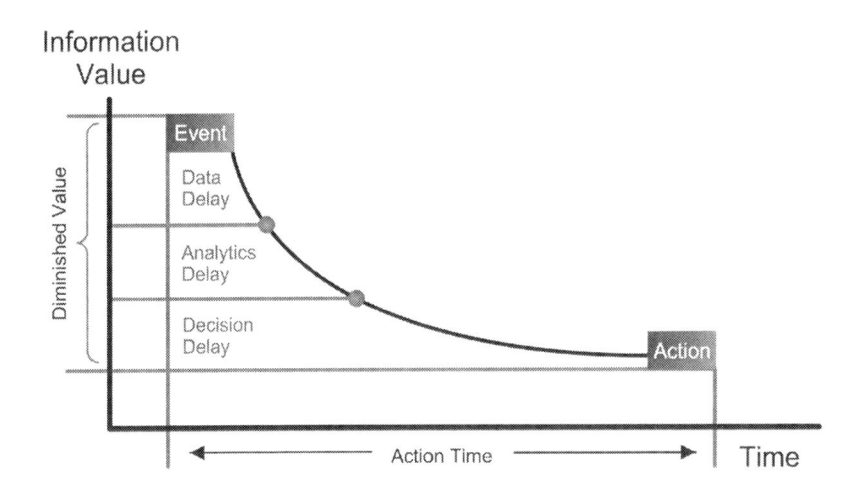

The traditional relationship between information receipt and action taken

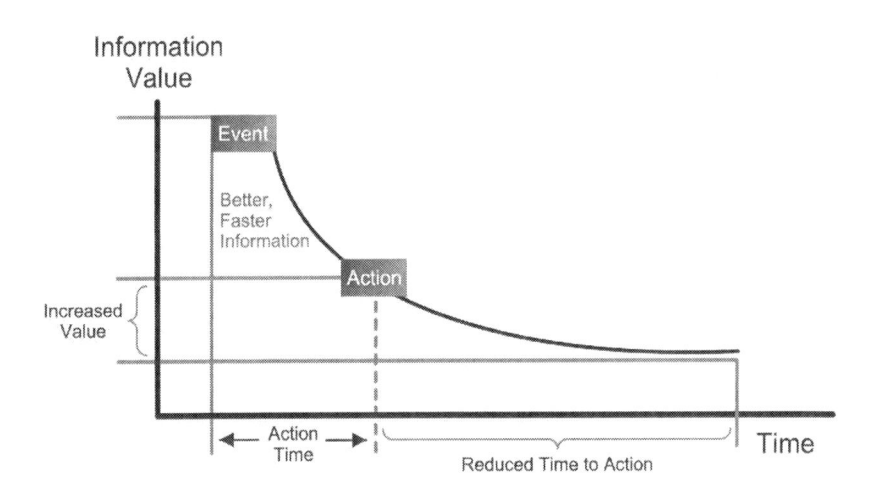

Decisions can be made faster when decision-makers get richer information, more quickly. Interoperable data enables trusted information to be shared in near-real time, supporting faster decisions based on more factors

Accurate information is a key weapon in the fight against uncertainty and indecision. Better, faster, more focused and accurate information, hopefully confirmed by more than one source, allows you to make a better decision, and it's far superior to doing nothing or acting on unverified data. You probably knew that, but we'll look at how interoperable data can help.

360-degree threats and 360-degree life. Have you ever experienced an abrupt shift from one context to another? Imagine sitting in a major strategic meeting in an office, with your laptop open, and getting an email telling you a close relative just died, or that your child has been injured in a schoolyard accident. Your context shifts immediately. Global events such as 9/11 cause the entire world to stop and shift – they're a major shock to the system of perceived normalcy that we've built up.

These kinds of mass-impact moments have always occurred (the assassination of U.S. President John F. Kennedy, for example) but prior to the world-wide instantaneous news we have now, information was filtered first to news organizations (radio, TV, newspapers, etc.) and then to the public, sometimes country by country. When you got the information depended on where you were; it might reach you hours or days after it happened.

The world's largest-ever recorded earthquake (9.5 on the Richter scale) occurred in Chile on May 22, 1960, causing a tsunami and killing up to 6,000 people. News went out via the newspaper, radio and TV; some people may have not learned about it for days. The world's second-largest ever recorded earthquake (9.3) occurred on December 26, 2004, in the Indian Ocean, and the world watched via cell phone cameras in real time, as it happened. The speed of the world's communications is accelerating at greater and greater speeds. This bears repeating because it will have far-reaching effects on today's technological evolution, and in order to be the Wayne Gretsky of interoperable data, you'll need to watch how fast the "puck" is accelerating, and plan on meeting it. (Wayne Gretsky is a retired hockey player who famously credited his success to skating to where the hockey puck was going to be, instead of where it was).

Change is the only constant. Today, everything is in motion and requires continual reassessment. In the financial world of 2009-10, corporations have been failing and scrambling for government bailouts. You can watch this happening in real time, on the web or on any one of 200 digital high-definition channels. All it takes is one bad report, and 20 minutes later the stock market is down 500 points. The rest of the world is becoming equally fluid and dynamic.

Government issues. While getting the right information at the right time is a formidable problem for individuals and organizations, it's particularly difficult for governments. In the U.S., for example, from the local town to the federal bureaucracy, there are at least 89,000 different jurisdictions. The U.S. government's reality of information sharing includes connecting federal, state, local, tribal, international, public, and private interactions (FSLTIPP). The failures of 9/11 vividly illustrated that the U.S. intelligence community was fragmented and disconnected from its operational capability. Little has been done to rectify this problem, despite billions of dollars spent.

This lack of progress is not for lack of effort and good intentions, but (as numerous blue-ribbon commissions and audit groups have proclaimed) there are few identifiable, usable, or applied end results. It is an intractable problem, and will take major efforts to move forward. Interoperable data can help make this work, as the cultural barriers are overcome.

First responders have been focused on radio, trying hard to get more interoperability from a platform that has faithfully served police, fire and other emergency responders over the last 50 years. Perhaps if we keep investing it will all work out, right? In reality, we need a new set of ways to get things done, but it's always tough to replace a method already in place with an evolved alternative without a forcing mechanism.

You'll remember that in the immediate days after 9/11, the U.S. was able to get a substantial number of things done because everyone was focused on the solution – not the problems, or the turf, the objections, or the cost. People and communities can respond to acute events cohesively. But for chronic problems, the response too often fragments into bipartisan bickering and political maneuvering.

Major issues surrounding interoperable data

Some of the biggest issues have little or nothing to do with the actual technology and data transactions, but are important to understand for a total context around the give and take of interoperable data.

1. **Culture.** Some organizations are just not oriented to sharing. Many decision-makers in the U.S. intelligence community and corporate America have been trained in a Cold War-spawned, "need to know" mentality. Despite the momentum generated after the failure to connect the dots around the 9/11 attacks, there's been little movement toward active, electronic sharing. (New government information sharing campaigns with catchy slogans, e.g., "Dare to Share" and "Responsibility to Provide" have been ineffectual, in my opinion). Sharing is hard, but the government is making it too hard and complicated.

2. **Policy.** Rules and policies governing the exchange of data can range from the simple to the sublimely complicated. Publicly available data might contain a simple disclaimer, while "sensitive but unclassified" information might be subject to a vast and confusing set of restrictions on use, reuse, safeguards, further dissemination, revocation, and a host of other complicating factors. A corporation might have sharing policies it enforces with supply chain partners, but applies loosely across the company. The rules for sharing and exchange are important to get right, and apply consistently.

3. **Authentication.** If you receive a set of interoperable data, how can you be sure it's authentic? What if the source had somehow been compromised and erroneous information integrated, which you then incorporated into your system, into future decision-making matrices, and exchanged with other entities – all in good faith? In certain cases authentication can be assumed, but in other cases it must go through a high assurance process, perhaps multi-factor authentication, encrypted data exchange, and/or other safeguards to ensure that the sender and receiver are both satisfied with the authenticity of the exchange. A thorough risk analysis of sources, and of the impact unreliable sources could have on your capabilities, is a primary element of establishing an information exchange.

4. **Privacy.** Personally Identifiable Information (PII) is any data element that ties a transaction to an individual. Examples are your Social Security number, passport number, name, address, and telephone number. Any of these (and there are many others) can turn an innocent piece of data into something that can compromise an individual's privacy rights. To

complicate matters, the rules defining and governing privacy rights can be vastly different on the sending and receiving ends of a single transaction. Imagine a passenger manifest list for an airline, shared across international boundaries with different airlines (for connections), commercial partners (for catering), and security organizations (terrorist watch list screening), and you can see a few of the issues that could come into play from a privacy standpoint. Sometimes the urgency of information overrides the privacy considerations of personal information. A good example of this is an Amber Alert for a missing child. In such a case, all kinds of PII may be released to the general public as quickly as possible in order to try to locate the child and ensure her safety.

5. **Timeliness.** Information can be originated, disseminated, and then changed or expired all in a moment, but also might last a lifetime. Knowing a storm passed three days ago would be irrelevant to most people, but a meteorologist might choose to aggregate that information into a much larger picture of climate change and fine-grained weather models.

6. **Sensitivity.** Information can be sensitive in a number of different ways, which can affect the way it is handled and disseminated. Over the years various federal agencies have implemented different (and often conflicting) policy and technology approaches. The result has been a massive hairball of confusion, which has resulted in reduced sharing, and also in people ignoring the inconsistent, confusing rules. The GAO (General Accounting Office, a key watchdog of government agencies) did a study that found over 100 different combinations of document markings and policies, just for the "federal sensitive information" classification.

Chapter One

Imagine being a city mayor and receiving three different federal documents marked "For Official Use Only" (FOUO) – you would expect to handle them all the same, right? In reality, if the three documents were from different federal agencies (for example, Department of Defense, Department of Energy and Department of Homeland Security), the rules for redistribution and destruction would be quite different. Help may be on the way:

> **Controlled Unclassified Information (CUI)** is a proposed government-wide framework that should, if implemented as designed, simplify and rationalize the federal government's approach to sensitive information controls. It would reduce the myriad of old classification methods to three simple classifications, and those classifications would be standard for all federal government agencies. This approach is very well thought out, and should have tremendous benefits in terms of streamlining government information flow over the coming years.

7. **Status.** Knowing the status of a piece of information (e.g., "current," "expired," "cancelled," etc.) is related to its timeliness. If a tornado warning has been issued, it is important to know the actual status – did the tornado happen, has the warning expired, or is it still an active warning? A detailed explanation of status management should be reserved for another book, but it is particularly important in data moving in near-real time, and across many organizations.

8. **Specificity.** The granularity of information is very important to its context, timeliness and status. Is this

message a general area notification, such as a thunderstorm advisory for northern Idaho? Or is it a report of an actual event that happened at 2:25 p.m., Tuesday, March 9, 2010, at 127 Elm Street in Portland, Oregon? These specifics can have a significant effect on the way information is treated and processed.

9. **Discoverability.** One choice that organizations, especially large ones, must make is how available to make information to other parties. If we can connect the dots, then so can those who mean us harm. Nonetheless, interoperable data gives the option for a high degree of discoverability.

10. **Assurance.** The relative correctness of the information that is subject to organizing and sharing is important. In a real-time space, speed may trump accuracy, but all organizations need to evaluate a strategy for mitigating the risk of bad information polluting the good.

11. **Semantics.** Information must be understandable and consumable by the recipient, and the message must be properly conveyed by the originator. For complex messages, this can be a tremendous challenge. Imagine a situation in which multiple manufacturers and distributors provide the same exact part, but each manufacturer gives the part a different number and description, in various languages and measurement systems.

It is also important to normalize the semantics when you combine messages from multiple sources. In particular, the definition of "when" and "where" can be particularly hard to discern. Is a time local or Greenwich Mean Time? Military time, or a.m./p.m.-based? Has the time been adjusted for the International Date Line? Latitude

Chapter One

and longitude give you a point on the earth's surface, but what if you are exchanging information on a submarine and its depth is important? It is less an issue if you are sharing data across town, but when information is consumed outside its initial context, these issues can become critical.

About the journey – this is not a joy ride

I am presenting a tool to you. As with many tools, there are proper ways to prepare, use, maintain, and improve these tools. This is a very sharp knife – ensure that you know what you want to cut, doubly ensure that you are grasping the handle instead of the blade, and that your other hand is protected as well. Implementing an approach strategy for interoperable data will have many challenges, and be different for each organization.

If your information can be turned on you and used to hurt your organization, you'll have to exercise more care than other organizations. An example is the Department of Defense, a very early pioneer with interoperable data. The organization is charged with creating, protecting, defending, transforming, and providing information internally and with other equally vulnerable partners.

The DoD is enormous, and the risk of information being used against it is very high. To counter this risk, a very complex implementation will result, with high levels of security, information assurance, and source authentication. In-transit encryption will be necessary.

For another example (just one of a plethora), consider the U.S. health care system in its current state of disarray. Everyone has electronic data, yet wherever I go, here I am with a pen and a clipboard, filling out paper forms. When I ask for my records in electronic form to take with me, I'm met with a blank stare. Policy makers need help from technical people, technical people need help from policy makers, and everyone in all the silos has to communicate with each other if this massive data dumping ground is to be organized and optimized. The lack of integrated information is dangerous, particularly for people whose care is coordinated (or not) among multiple health care providers, and those who have multiple prescriptions for drugs from various doctors. There are four million medical professionals, many of them independent business owners, and thousands of hospitals; getting them all to share information might require a small miracle.

The fact that the road is hard and long shouldn't dissuade anyone from starting. The first round of buildings that civilization worked on didn't stand the test of time very well either. We owe it to our collective future generations to move forward, harness our collective intelligence, and make improvements.

About transparency – government and industry

During Barack Obama's campaign and early in his elected presidency, the President declared that his administration would be the most transparent ever. This translates into sharing the behind-the-scenes working of government openly and broadly with constituents. The goal is to acknowledge the citizenry's right to know what is happening and provide information in time for that information to be accessed, consumed, analyzed, and evaluated, versus getting it years later, when there may be no recourse or way to change a course of action.

Chapter One

The Obama administration set up www.data.gov and other sites to make information broadly available. They've hired dedicated people committed to making information available at the raw data level. Raw data allows consumers of that information to draw their own conclusions, and apply a variety of tools depending on the analysis desired. For example, the administration intends to show how the entire stimulus package was distributed, from the federal government to the states to prime contractors, subcontractors, and grant recipients. This should result in independent analysis of expenditures, results, and jobs created.

Interoperable data plays a significant role in this process. Having information available in a standard format allows for uniform access and a common set of understandings about the information provided. It will allow for updates and improvements, and the blending of information from different sources. Done well, it will improve the distribution speed and enable multiple methods of analysis.

This is a huge challenge for the federal and other governments, but one worth pursuing, with a high payoff for multiple constituents. The more information is shared, processed, and revealed quickly and electronically, the better the feedback loop from multiple sources will be, and the more citizen watchdogs should be able to hold government accountable. NIEM.gov, the web site of the National Information Exchange Model, has a comprehensive set of information on interoperable formats that has evolved over the last several years, and will continue to be on the leading edge going forward.

Security note. I'm a big believer that sensitive, secret, or top secret information should be kept that way. It galls me to no end to see information shared that could cause our troops to get hurt or our intelligence gathering methods compromised, or give our enemies insight into our leading-edge technologies. I'm in no way recommending a free-for-all on information that is legitimately sensitive, and I would support a much harsher set of penalties for violations in this area. By the same token, we're fighting each other by not connecting information that could save lives, and that's just as frustrating. Recently, I watched with absolute shock how the information on the Fort Hood shooter was not shared, not connected, and not used to possibly prevent or mitigate this tragedy.

Large corporations need to share information. Transparency doesn't apply just to the federal government. Large corporations have been required to report results to the Securities and Exchange Commission (SEC) for years, but this information has often been unstructured and very non-standardized in format and usability.

XBRL (Extensible Business Reporting Language) is becoming the standard for financial reporting. XBRL is a set of standard, interoperable data formats that will allow electronic filing and a standardized approach to reporting financial information. This information, in turn, can be made available almost immediately to other groups, such as financial analysts, brokerage firms, investment advisors, and the general public. The standardized approach should allow for strong comparisons between like companies, and a more open and transparent understanding of a company's income statement, balance sheet, and cash flows. This is particularly important for publicly traded companies, but will likely become the standard for privately held companies as well.

There are many other industries where interoperable data reporting will allow for greater transparency and increased understanding. The electrical grid, which is very primitive in terms of technology (and under intense pressure to incorporate alternative energy sources such as wind and solar), will eventually be replaced by a "Smart Grid," a much more sophisticated means for electricity generation, management, transmission, and consumption. At the heart of the Smart Grid will be interoperable, standardized data elements that will allow multiple utilities, generators, customers, and distributors to share information with each other – in some cases securely, and in others, openly and transparently.

Democratizing Data, a book by David W. Stephenson, due out in summer 2010, outlines many of the issues and challenges of data transparency. I highly recommend it to anyone interested in understanding how open, transparent information will help the federal government be more effective, nimble, and forthcoming with information that rightly belongs to the country's citizens.

Chapter One summary

1. Interoperable data represents a large opportunity for organizations to be more effective and more efficient in their overall mission.

2. There are many issues within organizations that impact the sharing of information, and most are not technical. Being cognizant of these pitfalls will help avoid delays and shortfalls.

3. Interoperable data opens up the possibilities of shared data between organizations on a scale that dwarfs today's exchange of information in both breadth and speed of dissemination.

4. Interoperable data will help the government share data more quickly, and enable raw data to be available much faster, and in a format that can be used for many different types of analysis and purposes.

5. Transparency is an important element of trust and openness, and interoperable data will help this effort greatly when coupled with effective policies.

6. The world is rapidly reaching a tipping point in terms of the use of shared, structured interoperable data.

7. The destination and outcomes are worth the investment in time and technology.

An Opinion

The top ten benefits resulting from the broad adoption of interoperable data

....

Here's how a standardized, interoperable approach will help your organization now, and in the long term.

1. **Cross-organizational information sharing.** Within one organization, sharing information can be challenging, but the number of variables are reduced greatly, and responsibility can usually be traced to a single individual. When disparate organizations use interoperable data to move information among each other, the rewards include greater predictability, and enhanced views and understanding of particular topics. Sharing across organizational boundaries has risk associated with it, and needs to be carefully managed. The list of risks is long, but legal (violating your privacy policy), competitive (sharing your best customers), and inadvertent disclosure are three good examples. XML can help enable the exchange, mitigate the risks, and provide a strong audit trail of what was exchanged with whom, and when it happened.

2. **Fewer errors.** If the data you are processing is standardized, you will likely be able to set up strong controls on processing and incorporation of results. XML and other technology surrounding the exchange of XML files can insure that valid structures are passed. Logic checking can be incorporated, range checking on numbers implemented, and source integrity checks put into place to insure that the information was received from a valid source and conforms to valid processing rules.

3. **Service-oriented architecture** (SOA). Standardized data formats lend themselves well to being processed by automated services – without the need for a tight technical coupling of the underlying technology. Explaining the concept of a service can be easier using the non-computerized example of Federal Express: It is mostly irrelevant what is inside a standard Federal Express shipping container that has a viable set of delivery instructions attached, and has been properly paid for. This is of course within the bounds of rules and policies – shipping a live skunk would obviously cause downstream problems despite the standardized packaging! The weight might be bounded by a maximum as well. The packages can be picked up by couriers, dropped into shipping receptacles, sorted, packed, and delivered in a very standardized manner. So too can data transactions be managed by disparate computerized services. As long as the format and processing rules are all known, multiple systems can manage the information. An easy example is obtaining the latitude and longitude for a given address. Doing this inside your application is hard, and needs to be continually updated due to the complexity. By programmatically calling a centralized service provided by a specialized expertise provider doing this work for thousands of customers, the information can be obtained much more cheaply and have a higher degree of accuracy.
4. **Scalability and surge.** If you are processing uniform, standardized sets of data, you may well be able to move the information to an alternative processing capability such as an entirely different company facility or an outsourced managed service environment in the event that the amount of data vastly increases over time or suddenly surges due to an issue or problem. For many years, there were noticeable slowdowns on credit card processing around Christmas – the system was tightly configured for a certain

number of transactions, and due to the financial processing system limitations such as the number of connections and the number of simultaneous requests, could not be easily expanded when needed.

5. **Flexibility.** Standardized formats allow for multiple processing engines to utilize them, giving organizations options for using the same data for multiple purposes. NOAA (National Oceanic and Atmosphere Administration) has done a credible and impressive job of building an interoperable data interface to the vast collection of weather data that is continually being collected and analyzed by their impressive array of sensors, satellites and other collection vehicles. Rather than anticipate and build the innumerable web interfaces that consumers of the data might desire, they provide the data in a standard way and let the "customers" create their own results based on the raw data. Several investigations into Department of Homeland Security (DHS) and law enforcement sharing roadblocks have come to the same conclusion – "Give me the raw data faster and I'll figure it out" as the preferred paradigm versus delayed and sometimes irrelevant conclusions.

6. **Manageable change.** One of the advantages of using standardized formats is that changes are restricted to known, generational changes (for example, moving from version 1.1 to 1.2) versus an uncontrolled change that suddenly causes your system to fail (… and people to have to show up in the middle of the night to fix the mess). This is especially important in interoperable data feeds where predictable delivery and processing across many organizations are critical.

7. **Historical review.** Standardized transactions give you a means for storing, managing and retrieving information at future points in time. While proprietary formats may last, there is typically a dearth

of documentation surrounding them, multiple inline changes that have occurred, and general degradation that can eliminate significant value as time passes. Once, while I was working for a large software company, I built a large database or collection of customer registration cards that had been accumulating over the years and manually entered. Problem was, beyond the serial number, name and address, all the cards had varying marketing questions attached to them that quickly became impossible to quantify and correlate – and the company could not fully utilize a large amount of valuable historical marketing information. The main data was fine (address, name, serial number), but the rest became totally degraded.

8. **Unanticipated exchange.** The world is changing rapidly, and information systems that use standardized information sharing formats can adapt more easily to exchanging with new and unanticipated partners. In previous days, a sudden new data exchange requirement might require months of highly specialized interface development. With interoperable data formats that are well understood, interfaces can be enabled (rather than built) much more quickly. Rather than "welding" systems together, think of a solution where you can interlock Lego blocks, and you will get the magnitude of difference between proprietary interfaces and predefined, interoperable transactions.

9. **Trained and knowledgeable help.** If you utilize a standards approach, there will be many individuals and organizations available who can help you if you need extra capacity (surge), or to mitigate a key employee's departure, or handle a major incident/disaster. On the other hand, if your technical team goes off and builds something "special", getting any kind of outside assessment or assistance becomes very difficult. When standard building blocks are used, help can come from multiple sources.

An Opinion

10. **Strong thought process.** As interoperable data standards are designed, developed and approved, there is more scrutiny, analysis, debate, and coordination on the elements that make up the standardized format, across a much wider range of potential consumers of the data. Multiple subject matter experts are typically involved to insure that the functional design of the message is strong and useful across organizations, and technical lessons-learned can be incorporated into the design. You'll reap the benefits of the best thinking of a smart, cooperative community.

Chapter Two

Long-ago standards: things you'll recognize

Those who forget the past are doomed to repeat it.
—George Santayana

Santayana's quote has been pressed into service under many circumstances. In this case, if we review history we can find parallels that help simplify and explain today's complex environment. We can see how defining and adopting standards has solved many a sticky problem for the generations before us… and how we will have to re-learn the lessons about standards if we choose to adopt proprietary or non-standard methods in our haste to get something done.

Standards have helped the world move forward rapidly, especially when the standards had an impact on a large group of people or a significant industry. Much of what we take for granted today is the result of a standardized *process* that was accepted and incorporated into modern civilization. We have missed consolidation on a few – you've perhaps travelled abroad only to find a frustrating set of electrical plugs and voltages that varied by country. And if you've had the pleasure of driving on the opposite side of the road in England or Australia (versus the right or "wrong" side), you know what happens when a standard you've taken for granted since learning how to drive in the U.S. suddenly becomes a terrifying, hard-to-navigate, ordeal. Steven Wright summed it up perfectly when he jokingly mused, "Why is the alphabet in that order – is it because of that song?" Sometimes it is hard to know where the standard stops and the singing begins.

So here are a few examples of now-common standards to reflect on, in somewhat chronological order. Each has been hopelessly simplified in order to fit within this chapter, but you'll get the picture of progress through standardization.

The Great Wall of China. Picture the Great Wall of China – almost everyone will instantly get a mental recall of the third generation of "Great Wall" technology. What? Third generation? If you research the creation of the wall, you find that the first generation of the wall involved logs with dirt inside. This technology was effective in limited engagements, but hard to expand upward and outward, and subject to the elements – "Hey, the rains took out the wall again!" Maybe this generation was considered a "good" wall, but clearly was not in the "great" category. The second generation of the wall used field stones gathered around the project site. This was a much better solution – weather resistant, stronger, more extensible! Problem was, the stones had to be cemented by craftsman, and random-sized stones give inconsistent strength. And the bigger the vision and scale, the harder and harder it becomes to find the proper stones close by. So if you look at the wall today, you'll see interoperable building technology at work: uniformly sized building elements (bricks and carved stone) that allow for size and scale, variable production methods based on the local clays or rock quarries, and a lasting engineering tribute that can be seen from the moon. The remaining effort is truly a superlative accomplishment, ultimately enabled by an interoperable approach to construction.

Chapter Two

Weights and measures. When was the last time that you paid attention to whether something you bought was weighed or measured correctly? As soon as mankind started trading salt for shells and cloth for gold pieces, coming to an agreement on "how much" for "how many" became important. And once structures evolved past rude huts, standardized measurements of distance became necessary.

Early measurements were variable. A foot could vary by as much as three to four inches; a yard was the distance from the tip of the nose to the end of an outstretched arm. Cubits, spans, and other approximate distances made up the world of measurement for thousands of years. Weight was just as variable. The Babylonians used various standardized stones for different categories of weight; the butcher, horseman, wool-seller or fishmonger might each use stones of different weights. The Egyptians and Greeks used the wheat seed as their smallest unit of weight. The Arabs used a small bean called a karob as a means to weigh precious jewels; this standard has evolved into the "carat" used for diamonds and other gems.

The Romans inevitably intermixed many of these early standards as their empire grew and expanded over much of the known world. Being able to understand and discuss long distances became important, and scaled maps began to evolve. When the Roman Empire collapsed and Europe drifted into the Dark Ages, innovation in this area (as in so many others) was squelched. Sometime after the Magna Carta was signed, King Edward I of England set a permanent standard for the yard, which is very close to today's standard. England later revised the yard based on a uniform pendulum measurement.

Much later, in 1793, Napoleonic French scientists invented the metric system, which used the decimal system and brought order to the world of weights and measures. It was based on the meter – with

10 million meters representing the distance from the equator to the North Pole – a scientific measurement that could be validated. (After 17 years of implementation, France dropped the metric system, then returned to it for good in 1837.) Other standards followed, but adjustments in both metric and Imperial (feet-and-inches) systems continued. As recently as 1959, the length of the International Yard and the International Foot were agreed on in the U.S. and Commonwealth countries. The new lengths were shorter than the previous U.S. definition and longer than the previous U.K. definition.

Today, the modern metric system is used almost worldwide, with the United States the major exception. The U.S. toyed with the idea of adopting the metric system since before France adopted it the first time – Thomas Jefferson did a report calling for an advanced set of weights and measures in 1790. John Quincy Adams did another report in 1821, as America watched the rest of the world move toward the metric system. I remember learning the metric system in fifth grade and being told that the U.S. was going to convert. So much for that effort. It seems to be stalled for the foreseeable future, perhaps for lack of interest.

Interchangeable parts. Eli Whitney was one of America's foremost inventors. He transformed the cotton industry in the southern United States by developing the cotton "gin" (short for "engine"), which could automatically separate seeds from the white fibers that had come into huge demand worldwide – increasing production and speed to market dramatically. This had previously been a time-consuming, labor-intensive process. In fact, the design of the cotton gin was very simple and repeatable, which was a huge frustration for Whitney. Despite the patent he was granted for his device, everyone copied his design at will, causing him to move north to Washington DC.

Chapter Two

His real contribution to interoperability came after he returned north. At that time every gun was crafted by hand, limiting the size and scale of the military forces. Eli received a contract from the government to produce 1,000 muskets from interchangeable parts – an attempt to circumvent the bottleneck that limited the production of firearms to the number of qualified gunsmiths. History shows that Eli accomplished his goal, and that interchangeable manufacturing's vision was ratified. Parts from one firearm could be interchanged with parts from a like firearm – a major accomplishment that allowed a significant increase in manufacturing capabilities of new rifles, and, just as importantly, facilitated field maintenance and the combination of parts.

I heard a follow-on comment to Eli's genius at a lecture about object-oriented programming in Sausalito, California. After much hard work and experimentation under his government contract, Eli was apparently over budget and behind schedule, and the government was going to cancel his contact, bankrupting his company. The story goes that in recent times, someone went back to the original parts that were still in the Smithsonian and discovered microscopic file marks – Eli had faked the demonstration!!

Eli Whitney's technology later achieved very large scale at Harper's Ferry Armory, site of John Brown's uprising in 1859. The armory's mass production techniques (interoperable standards at work) were so impressive that Robert E. Lee, commander of the Union forces that put down the John Brown uprising, made it one of his first priorities to capture and move the technology south when Virginia and other states seceded and formed the Confederacy – enlisting Lee as their commander.

Railroads. Most of the world's railroads today (and virtually all in the United States) run on standard gauge tracks: 4' 8½" between the inside edges. Getting to a standard track size took a long time, and in some cases, having a differently sized track was considered a means of maintaining the status quo, or even a competitive advantage; manual transfers at a border from one train to another generated jobs, tariffs, and other transitory items. Economies of manufacturing scale and interoperability across states and countries eventually moved everyone toward a common standard. Once a common standard was in place, cross-border transportation systems could be built, and the flow of commerce increased greatly.

There can be downsides to the unfettered movement of resources. The Nazis in World War II were able to repurpose captured railroads in Poland and other countries from peacetime conveyances into key wartime assets because of the interoperability of the train tracks between countries. Progress can have a price.

7.62mm NATO ammunition. The history behind the development of this standardized ammunition could fill an entire book. For our purposes, think about seven different men from seven different countries in a World War I foxhole, each with a different rifle and a different type of proprietary ammunition. No matter what happens, there is no interoperability between them – if one has 5,000 bullets, it does none of the others any good. When NATO standardized on the 7.62mm cartridge, suddenly everyone could make guns and bullets that worked together – a German gun could shoot French bullets, an American rifle could load Italian-made ammo, and interoperability was achieved.

I've been told that the Russians designed their 7.62mm cartridge just a bit bigger than NATO's so that Soviet bloc countries received two standardized advantages: interoperability between Soviet Bloc manufacturers, and the added bonus of using NATO ammunition without reciprocating that capability to the NATO Alliance. The slightly larger cartridge would jam NATO rifles, but the Soviets could use either – making captured ammunition a real prize!

Shipping containers. As a child, I went to a replica of a colonial village somewhere in New York State. Beyond the butter churning demonstration, the hog pit, and other depictions of frontier life was a ship at the docks. Dock workers rolled barrels, lifted boxes, and carried crates on and off of the ship – all for show, but it made the point. This laborious "break bulk" process – hand loading and unloading of ships – was prone to damage, errors, and shrinkage. It continued until 1956, when shipping started slowly moving to an interoperable standard that changed the civilized world as we know it: the uniform industry shipping container that has standardized to a metal box 40' long by 8' wide by 8½' high.

Malcolm McLean sent the first 58 containers from New Jersey to Texas, initiating one of the biggest industrial changes the world had experienced. In less than 40 years, dockworker productivity increased by over a factor of 8,000, and seaports rose and fell in importance based on their adoption of this rapidly emerging new standard. Ships were subsequently built around the container standard, allowing ship speeds to increase, and prices for international movement of goods decreased dramatically. An entire industry grew up around the container business – loaders, truck beds, and innumerable other devices were needed – all built around the basic shipping container standard. Interoperability with the trucking industry helped the railroads flourish.

Why did it take so long to figure this out, when this seems so apparent in hindsight? What an amazing transition, all in a relatively short period of time. A triumph of standardization and interoperability!

The Uniform Commercial Code. Not all of the standards that make civilization great have been based in hardware and physical devices. Standardization across multiple legal jurisdictions can be illustrated by the adoption of the Uniform Commercial Code (UCC). There was plenty of commerce taking place in the United States during the 1950s, but varying laws, customs and other elements of commerce were grinding together as the world moved from buying and selling locally toward the expanding global marketplace we have today. If sales transactions were different from one state to another in today's world, imagine the limitations placed on companies like Walmart, McDonald's, and other entities that rely on a reasonable set of predictable rules and regulations.

Ecommerce on the Internet spans the globe, and often you don't know where the actual transaction is executed. This can be problematic, as illustrated by recent off-shore irregularities in online poker playing: Several individuals were caught cheating and making an inordinate amount of money by knowing all the cards that other players had while playing. No UCC application here!

Getting all the states to adopt the UCC was a laborious process, and minor variations still exist today. In general, however, the creators of the UCC did the country a great service by making the rules the same across a wide swath of America.

Chapter Two summary

1. The world moved forward because of standardized approaches, and we take many of them for granted today.

2. Often chaos and conflicting approaches have been the norm when a new capability emerged, but gradually (or abruptly) competing versions coalesced into a standard.

3. Standards can outlive their usefulness – how many telegrams have you sent recently? The fax has gone from a novelty to a standard tool to rare usage, all in the period of 30 years.

4. Because of our shrinking world, standards are going global at a faster rate than in the past, and are required to keep the wheels of global commerce moving.

5. Standards come in many flavors and colors, and a poor standard (e.g., inconsistent standards in electricity distribution across the globe, resulting in varying voltages and plugs) can cause a significant amount of frustration and loss of effectiveness.

6. Standards can enable rapid uptake of a new technology and change industries from the ground up, as we saw in the shipping container example – 40 short years to a total change in the way a major industry operated.

7. The military has begat many standardized approaches. Our examples included China's Great Wall, and interchangeable parts and ammunition; there are many others, including today's Internet Protocol, initially conceived and designed as a redundant means of messaging in case of a nuclear attack.

8. Non-technical examples provide a good reference point for implementing an interoperable data strategy, and can ease explanations to groups that are dealing with setting standards.

Case Study

Jefferson Parish, Louisiana tackles Hurricane Gustav

....

The following case study illustrates a standards-based approach to emergency management. The services used incorporated physical and electronic information standards to enable an off-the-shelf, multi-channel, rapid response communications solution that supported timely, effective response in an area devastated by a hurricane.

Challenge
Approaching Labor Day weekend 2008, Jefferson Parish public safety officials faced the threat of yet another hurricane pounding ashore and wiping out roadways, power lines, and vital communications networks. After Hurricane Gustav struck, how would they assess damage, share critical needs, and coordinate recovery with responding local, federal, and state agencies as quickly as possible?

Solution
Parish leadership worked with CommsFirst, using an established CommsFirst managed services plan that included accessing response experts and rapidly deployable, self-sustaining communications solutions.

Results
A CommsFirst deployment team arrived just before Gustav made landfall. Over the next 96 hours, the team quickly established (and re-established) multiple ad hoc networks throughout the parish, adapting to the evolving demands of the incident. Among the hardest hit was Grand Isle, a barrier island approximately 100 miles south of New Orleans. Officials there were completely isolated, without power, water, phones (cellular or land lines), or radio

communication. Bridges were impassable. The parish flew a CommsFirst expert who carried a CommsPack – a self-contained and self-powering backpack with integrated solar power, portable satellite antenna, Wi-Fi router, and VoIP and satellite phone capabilities – to the island via helicopter. The CommsPack enabled:

- Establishment of an incident command post within minutes of arrival
- Rapid damage assessment and data sharing
- Phone calls and Internet access for the mayor, police and fire departments, EMS, city council, local business owners, and others stranded on the island
- Connection with Governor Jindal, Jefferson Parish leadership, Homeland Security, National Guard, and others to coordinate recovery
- Interoperability for P25, 700 MHz, and 800 MHz radios brought to the scene

After the bridges were clear, a CommsFirst Operations Vehicle (OP-V) arrived from the parish 9-1-1 center (after relieving call congestion there), and a second soon followed. Together the OP-Vs provided two secure Wi-Fi hotspots (one-mile radius each), 26 phone lines, streaming video, and radio interoperability for local officials at city hall and disparate responding agencies arriving at the incident command post.

By the time a responding command bus was up and running – an hour after cellular and push-to-talk service came back on line late Friday – the parish had been making calls, sharing data, and coordinating their recovery efforts for well over 96 hours.

Chapter Three

Near-past lessons: standardized technology and computing

> *All parts should go together without forcing. You must remember that the parts you are reassembling were disassembled by you. Therefore, if you can't get them together again, there must be a reason. By all means, do not use a hammer.*
> —IBM Manual, 1925

Now we'll move forward into the world of computer technology, looking at a high-level picture of an industry that has moved fast and adopted many standardized approaches, consequently making historic achievements in a very short time frame. Reviewing overall trends and understanding some of these approaches will lay the foundation for looking at standardized, interoperable data.

The first electronic computers – amazing machines for their time – showed up in the early 1940s; their development was accelerated by the World War II effort. These early computers consumed massive amounts of power and took up lots of room, while providing incredibly small capability when compared to today's simplest machine. They were made with manual switches, vacuum tubes, and (prior to the invention of the integrated circuit) tiny metal "doughnuts" that could be set to a 1 or 0 through a three-wire system. This primitive equipment spawned the information age, with subsequent change, turmoil, and mind-blowing advances. Today a single iPhone's computing capability dwarfs all computing capacity – combined – that existed in the late forties.

In the early 1960s, my older brother Dave worked in finance and payroll while in the Army, and learned to be a "keypuncher" – creating standardized paper input streams of pay changes, etc. When his Army enlistment was over, he was hired by IBM in Poughkeepsie, New York, to do the same type of keypunching. IBM was in great need of computer programmers for their rapidly expanding base of electronic mainframe machines, so they gave Dave an aptitude test and trained him to program in IBM's assembly language. (He retired 30 years later, having contributed greatly to several world-changing projects.)

My first exposure to a computer came when I was eight, at an IBM Employee Open House in Poughkeepsie. I remember playing Tic-Tac-Toe with a light pencil against an IBM 360 class mainframe as a magical experience. Who would have guessed where things would go from there? Even Thomas Watson, one of the founders of IBM, initially thought that there was a worldwide market for only five computers.

Interoperability milestones on the information highway

1. **COBOL.** When computers were first developed, programming these behemoths was a task for very sophisticated individuals with deep math and engineering backgrounds. Much of the early work was done in binary machine language (literally

Chapter Three

typing 1s and 0s, or flipping binary switches on complicated control panels) and assembly language (one step advanced, but still very close to the machine architecture). Early applications were impressive (for the time). As an example, calculating artillery trajectory tables (which took weeks to do by hand with slide rules) could be accomplished in seconds – a huge time-saving application.

Rear Admiral Grace Hopper, a legendary computer scientist and Navy officer, developed a more English-like programming language called Flow-Matic. The Conference on Data Systems Languages (CODASYL), an industry consortium devoted to developing a standardized computing language, took Flow-Matic and turned the best of it, with other innovations, into the Common Business Oriented Language (COBOL). COBOL was adopted and became the basis for many millions of lines of computer code worldwide, with acceptance driven by IBM and other computer manufacturing companies.

COBOL was so long-lasting that in the run-up to the year 2000, when I needed a COBOL expert to help prepare my company's networked system for potential changes and problems, I couldn't find anyone to hire; there just weren't enough people still working who knew the language. I ended up contracting with a couple of programmers living in a retirement home in New Jersey – the language had outlived many of its developers!

COBOL was one of the first open-standard, interoperable programming languages that could work on different computers. While it never achieved true portability (computer vendors always found a way to add small extensions and proprietary items to their compilers in the effort to lock customers in), it was close, and set the stage for many innovations in the business and computing world.

2. **The IBM PC, MS-DOS, and Windows.** When the IBM personal computer (PC) burst into being in 1981, it was a defining moment for the computer industry and had impact around the world. What most people don't remember is that there was already a plethora of PCs and operating systems available at the time – the Apple II, Tandy's (Radio Shack) TRS 80, CPM machines, and the Xerox Star (which had most of the elements of the windows, icon, mouse, pointer [WIMP] interface initially popularized by the Apple Macintosh and years later by Microsoft Windows).

The IBM PC was a defining standard because IBM defied its own traditions and made it an open, standardized system. Within months of the IBM PC release, there were numerous compatible machines including the Compaq "luggable," a beast of a portable PC (at 28 lbs. and about $3,500) and a harbinger of what laptops would be. An entire industry standardized around Microsoft Disk Operating System (MS-DOS, released in 1982) and the open architecture – and millions upon millions of PCs were sold. Michael Dell built a billion dollar business out of the trunk of his car! Bill Gates had the vision to license MS-DOS to IBM and others on a per-copy basis instead of giving up all rights for a one-time development fee. This foresight resulted in his becoming the world's richest person. All this was due to the power of a standard – one that Microsoft happened to control, allowing the company to capture and hold nearly 100 percent market share for many years.

Chapter Three

There were many lawsuits against Microsoft for monopolistic behavior, but imagine how much more slowly the industry would have developed if MS-DOS (and later, Windows) had not served as the central operating system for the hundreds of millions of PCs that evolved during the 1980s and 1990s. The standardized structure made it safer for organizations to make investments in technology, and encouraged developers to come up with new applications. A 5MB hard drive was massive for a short period of months, quickly replaced by a 10MB, then a 30MB, then ever-larger drives. New generations of standardized processors (built mostly by Intel Corporation) also set a new performance bar almost every 18 months – the 286, the 386, the 486, and then the Pentium. People talked about the specifications in great detail; clock speeds could and would be debated at cocktail parties. Today, much of the newness has worn off, and few people know or care about the clock speed of their PC or Mac – the novel has become the commonplace.

3. **Packaged application software.** When standardized software applications, first VisiCalc and then Lotus 1-2-3, were built to serve the increasing base of personal computers, we saw the power of standards at work again. The standard machine enabled standard software to be sold in massive numbers. One notable screw-up here was Digital Equipment Corporation (DEC), at the time the second-largest computing company in the world. Having introduced a standardized VAX minicomputer architecture that allowed the same operating system to run across a wide range of machines (very standardized, and very forward-thinking for the time), DEC introduced an IBM-compatible computer of its own – the DEC Rainbow. However, DEC produced the Rainbow with a

different disk drive format and a couple of other non-standard "improvements" that basically killed its sales to everyone except captive corporate customers of DEC. DEC was later surpassed by multiple companies, and eventually purchased by Compaq and then Hewlett Packard – a minor rival in DEC's heyday of proprietary minicomputer solutions. DEC's proprietary advantage was definitely trumped by the standard and interoperable nature of the PC.

Hardware advances allowed packaged application providers to add more features, which in turn drove a demand for further improved hardware and a network effect developed from there. Hardware that functioned perfectly might be retired in order to run the latest version of packaged applications in attempts to increase productivity of the workforce. The hardware and software synergy produced a strong current of change.

4. **The Internet** took decades to develop but reached its tipping point in the early nineties, and things began to move very quickly. Tim Berners-Lee spearheaded the effort to create the information management system we now know as the World Wide Web, which launched in 1991 – leveraging the infrastructure of the Internet, which was far more developed. The WWW created a "web of hypertext documents," which could be linked to each other regardless of physical location. In 1993 the National Center for Supercomputing Applications (NCSA) launched the Mosaic web browser, and web use exploded. Marc Andreessen, leader of the Mosaic team at NCSA, started Netscape and released Netscape Navigator (which depended at least partly on standards established by Mosaic) in 1994. At its peak, Netscape accounted for 90% of all web use. Microsoft initiated the industry's first browser war by

Chapter Three

bundling Internet Explorer with Windows, and took over the browser market. As I write this, half a dozen browsers from other companies are challenging that dominance, including one from Google.

HTML, the lingua franca of the World Wide Web, was both a miracle and a curse of a standard – miraculous in that almost anyone could create these entities called web pages, and anyone with a browser and the correct Uniform Resource Locator (URL) could consume them from anywhere in the world regardless of the type of computer they used or network they were attached to. The curse was that the fairly immature standard took hold very quickly, and just as quickly ran into problems and irregularities that became harder to solve because of the rapidly expanding base of amateur programmers. There were tricks for each version of browser, and religious wars inside organizations fought over dumbing down the HTML or taking full advantage of proprietary techniques.

What many people also don't know or remember is that there were many standards working their way into the mainstream for years before the World Wide Web made Internet use commonplace: TCP/IP and DNS, to name just two. The Internet was built on the idea of interoperable data packets that could be disassembled, transported over various routes, and then magically re-assembled at the receiving location.

There were plenty of other networking protocols at the time – IBM had SNA (System Network Architecture) and if you were "true blue" (meaning every piece of your equipment was from IBM), you had a good chance at connecting everything together. Hewlett Packard and Digital Equipment Corporation both had networking protocols that connected their own equipment, but having multiple vendors' equipment on the same piece of cable or connected in the data center was a far leap. Many of the vendors disliked the idea of a unifying method for connecting – the fear (now realized in many environments) was that hardware would become a commodity, and cause the vendors to compete on price versus proprietary advantage.

As TCP/IP became a dominant standard (driven heavily by universities that couldn't afford separate networks or to go all- IBM), unified networks that could handle all types of computing equipment were increasingly favored. Proprietary standards were relegated to the back of the train and ultimately consumed by a single, general purpose standard.

5. **Personal computer networks.** In 1987 I wrote *The Computer Networking Book*. At the time, there were three major competing network standards for personal computers – Ethernet, IBM Token Ring, and ARCNET. ARCNET was aging, and IBM had a lot of proprietary juice behind Token Ring Networks. IBM touted the deterministic nature of their network as far superior for real business than the California-conceived and developed Ethernet. Ethernet was like a room full of rabid conversationalists – the first to get heard was serviced, with all others backing out and retrying until they got a slice of the network. But Ethernet was more open and highly adaptable for multiple kinds of hardware, and quickly evolved to become the de facto standard for all mixed computer environments.

Chapter Three

6. **The iPod and digital music.** One interesting standards story revolves around the digital music market. Music and the emerging web collided in a major way – music files were small, very shareable, and used repeatedly. An explosion of ventures all tried to gain the right set of ingredients: software, business models, and position with the record labels. The record labels dismissed the entire process and barely got involved, since CD sales continued to grow through the 1990s. Music pirating took off in a major way when Napster came on the scene – the company flaunted copyright laws, and created a zealous following of people getting music for free from each other through file sharing on the web.

Lawsuits ensued, ventures collapsed, and the digital music market was a scorched-earth market for future investment. CD sales also started shrinking as people realized that buying an album for one or two good songs wasn't worth it. Out of the ashes came Apple and the iPod, with a beautifully designed player at an affordable price, a means for acquiring music legally (iTunes), and an ongoing support structure that people trusted. Sales skyrocketed, even though the environment was entirely proprietary – there was such strong demand for a viable answer to this problem that nobody cared that there was only one source.

A strong competitor has yet to be seen for the iPod, though Apple has relaxed some of the restrictions, and is much more capable of dealing with the music industry than previous vendors were. Surely there will be many more evolutions in this market, but Apple did a great job of addressing demand and providing a standardized (though hardly interoperable) approach to the digital music market. There is a very good article on the "good enough generation" in a 2009 *Wired* magazine that makes the case that the high availability of low quality music files trumped high fidelity.

7. **USB chargers for cell phones.** Proprietary hardware generally provides more profit to the manufacturer and less convenience to the customer. This has certainly been true for cell phones, resulting in the need to travel with a charger (or risk a dead battery), and to purchase one or more new chargers with the acquisition of a new phone. Early in 2009, European mobile phone manufacturers agreed to a USB-based industry standard for a universal charger for new mobile phones, to be implemented by 2012, and U.S.-based companies are discussing the possibility of adopting the same standard. Such a move would not only be more convenient, but also save an estimated 51,000 tons of discarded phone chargers annually. You can probably think of many more examples like this one where a little bit of normalization goes a long way.

Chapter Three

Chapter Three summary

1. Computing is still a young field, coming into mainstream acceptance and use within just the last few decades. As an industry, we have much to learn and many generations of change still to come as information technology impacts every facet of our lives. Imagine the technology 100 years away, and 1000 years from now, and you'll get the idea.

2. Standardized approaches, though sometimes manifested as proprietary market ownership, have allowed for megagrowth within the computer industry – first in hardware, then software, and most recently the World Wide Web. Open, free interoperable data standards are the next logical step.

3. Ignoring an emerging standard can mean death for a business or technology in a rapidly moving marketplace. I had the poor fortune of launching a presentation graphics software product a month or two before Microsoft launched PowerPoint. We got crushed.

4. The combination of standardized networking, hardware, and a web interface via a browser has brought computing capabilities to billions today, and will enable billions more in the future. The whole world will be connected at light speed.

5. Demand can sometimes drive a standard, as the story of the Apple iPod illustrates. Interoperability can suffer if the standard is used to lock the customer into the proprietary environment.

6. Interoperability is important even at our basic life level – exchanging calendars, moving information from one device to another, and so on. It is both a global and personal opportunity for streamlining the time spent.

Chapter Four

The beginning of standardized data transactions

*The great thing in the world is not so much where we stand,
as in what direction we are moving.*
—Oliver Wendell Holmes

Often, mankind dreams up big ideas long before they can become reality. Around 1493, Leonardo da Vinci conceived of a machine for vertical flight; it was hundreds of years before available technology made it possible to realize his helicopter. So is the case with standardized data. Long before XML had a name (XML is a set of document-marking standards; more in the next chapter), it was a concept. Industry pioneers had a shared vision and made exploratory moves toward standardized data that laid much of the groundwork for today's more sophisticated ideas and easier-to-use technology.

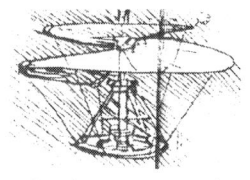

A 15th-century Leonardo da Vinci sketch of a vertical flying machine
Source: Wikipedia Commons

In this chapter, we're going to focus on examples that are data driven, building up to today's environmental opportunities and challenges. We'll look at punched card technology, the airline reservation system, the VISA network, and Electronic Data Interchange – all harbingers of today's developing XML and interoperable data standards. As with most innovations, early models have very limited use and very high costs, then gradually become more mainstream and affordable. As an example, around 1945, engineer Percy Spencer serendipitously stood in front of a magnetron (the power tube that drives a radar set) and noticed that the chocolate bar in his pocket had begun to melt (hopefully in his pocket protector, itself a new invention at the time). This led to the development of the "radar oven," an expensive, strategic industrial tool which matured into the commonplace, inexpensive microwave found in everyone's kitchen today.

The punched card

Dr. Herman Hollerith began the information processing revolution by designing a machine that could tabulate punched cards. At the time, punched cards were used in textile manufacturing, fed into looms for weaving patterns. They were also used for information analysis, but they had to be tabulated by hand. Hollerith took punched cards to a new level by applying a newly available technology – electricity.

On January 8, 1889, Hollerith received a patent for his breakthrough machine, one of the forerunners to modern computers. His advanced machine could read the information punched into holes in the cards. His performance on the 1890 census was impressive – in one year, he did work that had previously taken eight years to accomplish by hand, astonishing the government and setting the stage for one of the great American companies to develop. Hollerith's technology became the foundation for International Business Machines, now known as IBM and still a world leader in the information sharing revolution. If you enjoy history, you might delve into the history of the punched card more deeply. The stories are interesting, the details intricate, and the personalities fascinating; it's a good reminder of how complicated change always becomes by the time it matures.

Chapter Four

In many instances, it's the blending of one technology with another that generates the spark for a subsequent breakthrough and the next generation of capabilities. In Hollerith's case, it was electricity coupled with the card readers that made the quantum leap forward. In other cases, breakthroughs such as the transistor allowed the size and power of computers to increase exponentially without much change in form factor. Companies such as Intel just figured out how to pack more capability into the same-sized chip.

The airline reservation system

By the late 1950s, computer technology had been established and was increasingly used for specialized applications around the world. Automated payroll was a breakthrough business system; on the scientific side of the equation, tabulating engineering tolerances was an early success. The next step on everyone's mind was combining computerized systems with other major breakthroughs.

At the time, most applications were localized: run on one computer in one place, using massive batches of punched cards. Computer scientists began to dream of ways to minimize—or outright eliminate—ties to a single physical location. If you could figure out how to distribute applications separated by a few miles... then many miles became possible. Thus the network was born, eventually leading to our interconnected world.

In the 1950s, the American Airlines ticketing system was entirely manual. AA wanted a system that would allow real-time access to flight details around the world for all of its offices. With help from IBM, the Semi Automated Business Research Environment (SABRE) system was launched as a trial in 1960, and took over all booking functions by 1964. Through this system, one central repository for flight inventory was maintained and made available worldwide.

This capability required hard-wired terminals and substantial training for the operators who used the cryptic commands, but was an astounding new concept. It was eventually extended to include independent, third-party travel agents who then could also sell airline tickets. No real competitors got a foothold in the market. Systems for other airlines were developed using SABRE's framework, which had become the de facto industry standard.

Over time, the various systems merged, giving travel agents more flexibility to see different options for their customers. American Airlines, which controlled the SABRE system, was sued for manipulating the order in which these options appeared, and eventually had to offer even-handed access to all airlines and flights. If you are old enough to have experienced the use of paper tickets and a blue SABRE terminal at the travel agent's office, you probably appreciate the ease of eticket travel and the convenience of making your own reservations through services such as Expedia and Orbitz.

The impact of these worldwide systems should not be undervalued – they were truly global game changers and harbingers of things to come. While cumbersome and primitive by modern standards, these systems began the trend that is still accelerating today – moving all of man's knowledge closer to all of man. A few key strokes, and vast amounts of information are "at your fingertips."

The VISA network

One of the first end-to-end, standardized interoperable data networks was something we all take for granted today: the credit card processing network. It's an interesting and compelling story of inter-company cooperation enabled by interoperable data standards.

Chapter Four

Prior to credit cards, foreign travel was very complicated in terms of having enough cash, letters of credit, and contingency plans. Many times, a fat money belt was the only good answer – yet could easily make you the target for a robbery.

The VISA network evolved through standardized data transactions that could be passed from one participating bank to another. Guarantees were in place; security and fraud were scrutinized and managed. (Of course the bad guys have gotten more sophisticated as well, and security remains a challenge today.) We talked earlier about the cultural challenges to sharing information between organizations; apparently VISA encountered most if not all of them, including resistance, denial, negotiating, and ultimately acceptance and dominance. In a fascinating book, *One from Many – VISA and the Rise of the Chaordic Organization*, VISA founder Dee Hock describes the changes in thinking that had to occur among all the players to make money flow worldwide for consumers. The fact had to be accepted by all that no bank by itself could provide worldwide coverage, and in order for the whole to be greater than the sum of the parts, a new organizational model had to be put into place.

The heart of the network was the exchange and settlement process. VISA started out as a non-profit cooperative owned by all the member banks. There was a natural reason for everyone to cooperate; like the railroad containers, this new way of doing business revolutionized the travel industry, and enabled many more people to move freely about the world. The cooperative model also established a modicum of inter-bank trust – no one bank had control (and thereby leverage) over another, making it a more rational system.

When you consider the process, it's magical and straightforward at the same time. Pete O'Dell walks into a shop in Melbourne, Australia, where he is attending a conference on Internet Protocol Version 6. He purchases a gift and hands over his U.S.-issued credit card. Almost instantly, an efficient and standardized transaction is fully completed – name, time, location, amount (with currency exchange done in real time if necessary) – beamed through the network and analyzed, with a confirmation or denial of the transaction sent to the retailer. Upon confirmation, the goods change hands, the merchant is paid, the customer is charged, and all exit the transaction on a positive note. The transactions are virtually the same globally, so the customer can reconcile his monthly statement upon receipt, and the network allows for arbitration and dispute settlement.

Today, most credit card statements can be electronically ported into financial programs such as Quicken, allowing analysis of short- and long-term spending patterns, historical data, future projections, and other interesting information. The information can last for a very long time once standardized and stored electronically. You might not save 12 years of monthly paper statements, but it is very easy to save 12 years of data on your electronic personal financial system.

Electronic Data Interchange

We've just seen how interoperable, standardized transactions changed the world for the traveling consumer and international retailers. For large companies wishing to communicate with each other in terms of back office operations, the path has been a bit different. "Electronic data interchange" (EDI) refers to the structured transmission of data between organizations by electronic means. EDI was conceived and implemented as one of the first interoperable business networks, using standardization for computer-to-computer communications even prior to the Internet. The TDCC (Transportation Data Coordinating Committee) did groundbreaking work in the late 1960s around inter-company transactions.

Chapter Four

Early work by General Motors, Sears, and Kmart had spawned very effective, but proprietary, trading partner systems. If you were a vendor to all of these companies, you had three different methods of doing business, each with different protocols, security, authentication, and formats – and also proprietary hardware. This of course caused significant implementation and operating costs, and slower timeframes.

Over time, companies realized that a shared, standardized system was a better alternative for all. It could revolutionize industries, and supply chains could be longer and more reliable through this type of uniformity. Competitive advantage could be had in other ways, so the move was toward standardization and a degree of interoperability even among competitors.

Industry-specific EDI became commonplace, even though EDI was relatively difficult, expensive, and required the use of a third-party Value Added Network (VAN) partner in order to secure and validate the information. I worked with a small company in Joplin, Missouri, that was told by Caterpillar that in order to receive preferential vendor treatment, all purchase orders and shipments needed to be done via EDI. A simple ROI (based on the expected costs versus the added margin available to partners) made it an easy business decision to spend the $15,000 (a significant sum at the time) to develop and implement the capability. Purchase orders, shipments, and returns were all automated in a 180-day period, and an automatic interface developed between the Digital Equipment VAX manufacturing system and the IBM-based EDI system. Standardized transactions made all the difference.

Ingram Micro in California outstripped the competition in the PC computer distribution business because of a very standardized approach to process and interoperable data systems. CEO Chip

Lacey moved Ingram Micro from being a small player to owning THE center of the PC distribution business over a number of years – all while volumes were increasing, margins were shrinking, and products moving to worldwide markets. Staying competitive meant being able to handle a large increase in transactions at increasingly smaller processing costs.

Ingram Micro was also one of the first innovators in the XML world. In 1998 the company became an early driver of RosettaNet, an open, non-profit consortium, developing universal standards for sharing business information and ensuring that just one set of governance rules applies worldwide.

E-Z Pass

You've read comments in this book about how government is not very good at innovation and interoperability – all my opinion, of course. One notable exception, based on my experience to date and the large number of organizations involved, is the E-Z Pass toll system. If you live on the east coast of the United States, you likely have an E-Z Pass, or you've seen them if you've traveled here and noticed cars whizzing right through the toll booths. The system has different names in different places, but there are always identifying marks that let you know (except for Florida's Sun Pass) if you are interoperable using the E-Z Pass standard.

The magic of the system interoperability is the fact that many different states are participating, and I can use the same transponder seamlessly in all of them. I live in Virginia, but got my E-Z Pass device (a 915 MHz wireless transponder, if you were dying to know) from the state of Maryland. If the E-Z Pass balance gets low, it is automatically replenished from my credit card. The revenue

Chapter Four

collected has to be settled and accounted for just like the VISA network that we discussed earlier – among many members of one financial network. On a recent road trip from Alexandria, Virginia, to Seattle, I went through nine states with tolls and only one (Ohio) still had manual toll takers and was not integrated into the E-Z Pass association of participants. The designers and builders of the system between all the states should get great thanks for its functionality. The system undoubtedly saves millions of gallons of gas every year, countless consumer commuting hours, and captures the revenue very efficiently.

This technology could be a harbinger of automated shopping. Just go into a store, pick up what you want, and leave. There's no need for a cashier when all the items are tagged and charged to your securely identified account. No more getting home and finding out the kid inadvertently put a candy bar in his coat pocket.

Chapter Four summary

1. The vision of standardized data has been in the industry for many years; examples showed how standardization enabled major industry growth and connectivity. It was hard work with older and less capable technology, but had huge impact on the world. We've come a long way, baby.

2. Many of the lessons learned in these early implementations were brought forward into the current day by pioneers who are still active and involved today. This is one of the advantages of a young industry.

3. In some cases it was a single company that drove much of the innovation, forcing its suppliers to work with the standard the purchasing organization set, and achieving benefits that then pushed other companies in the industry toward a standardized, automated approach. Once a good standard is put into place, capabilities become more of a commodity and are assumed. Imagine a supermarket today without a scanning system at checkout.

4. Inter-company data sharing for back office purposes (inventory, purchase orders, shipments, returns, etc.) has paved the way for modern-day evolution to more sophisticated solutions that are moving to the consumer, and to increasingly capable smartphones.

5. The 1960s and 1970s brought systems that connected the world together in real time for important applications. You can see the influence of these applications in today's world in products available at much lower prices, and in much more standardized forms.

Case Study

Using interoperable data to improve diabetes management

....

MyCareTeam (MCT), a web-based diabetes management application developed at Georgetown University, facilitates diabetes management via information and communications technology.

Individuals with diabetes are required to test their blood sugar levels multiple times per day to maintain optimal glucose control. To do this, patients apply small drops of blood to a monitoring strip inserted into a glucose meter. The blood is analyzed inside the meter and the resulting blood sugar level is recorded onto a chip in that device. Most of the glucose meters on the market today provide a communication port that allows the meter to connect to a computer so that the blood sugar levels can be transferred to proprietary diabetes management systems. By connecting the glucose meter to a computer and using software like MCT, a patient can upload multiple blood sugar readings and other diabetes-related information to a secure database for analysis and review. The stored data is then available securely over the Internet to the healthcare provider, who can then treat the patient from afar and provide ongoing management of his or her disease.

At the same time, patients can improve their understanding of their diabetes and the impact of their actions on their blood sugar levels. This type of solution is less costly, more convenient, and more accessible than traditional diabetes care, which relies on office visits in combination with follow-up telephone calls between appointments to manage an individual patient.

Funded by the National Library of Medicine and the U.S. Army Medical Research and Materiel Command, and licensed to a commercial company through a technology transfer arrangement with Georgetown University, MCT can aggregate information from a wide array of blood glucose meters to a centralized database. These home monitoring devices use varying, non-standardized data formats. The MCT application serves as a kind of Rosetta stone, achieving interoperability by accepting various physical connections to the glucose meters (RS-232, USB, infrared) and permitting the extraction of data using various proprietary data and message formats.

Data from different devices are all normalized into a uniform record on MCT. This allows patients to use the meter of their choice, even from multiple vendors, and to upload their results to the secure web-based MCT system for a historical archive of their readings.

MCT has been tested in a wide range of clinical settings, cultural environments, and geographic areas. It has been shown to assist in the lowering of hemoglobin A1C levels (an estimate of blood sugar control over several months). MCT offers individuals with diabetes ubiquitous access to aggregated clinical data any time of day or night, to aid them in managing their health.

Chapter Five

The emergence of Extensible Markup Language; (XML); a tour of major initiatives

After growing wildly for years, the field of computing appears to be reaching its infancy.
—John Pierce

Moving into modern day

We've examined the footpaths, trails, and early roads that have led up to our present-day situation. This book has been published in 2010 for your reference. If you are somehow reading it in 2020, much of the cultural information will still be very relevant, but the technology will have moved forward at an accelerated rate. While we can transition through generations of technology very quickly, when it comes to people and cultures we're still dealing with Biblical issues. How many corporate initiatives have you observed where the solution was to divide the baby in two pieces? Have you seen the equivalent of Cain and Abel happening around your organization when the chips are down, and survival at a premium? I can guarantee that the technology will be more capable – but can't make the same claim for humanity.

In this chapter, we're going to move into technologies currently used for an interoperable data sharing approach. This won't be highly technical, though there are more TLAs (three letter acronyms) used than in previous chapters. If you don't have a technical background, focus on the high ground and stay out of the weeds in terms of details. The concepts are the keys to understanding how much easier interoperable data has become than in days past.

What is XML?

A concise definition: XML (Extensible Markup Language) is a standardized method for communicating data and definition between disparate computer systems. (There is much more to it, of course, but we're going to focus on what XML does for interoperable data, as opposed to what it is and how it's made. For more technical information, check the list of links, resources, and recommended books in the appendix.)

XML delivers the ability to define and share structured and repeatable data that is predictable, valid, and created in a way in which it is self-describing. This is an important point – if you have used other data exchange formats such as CSV (comma-separated values) when moving data from an Excel spreadsheet to a word file or a database program, you'll know that the person passing the information and the person receiving the information must know the meaning of each field – it is not usually contained in the data file. XML overcomes many of these limitations, allowing data to flow to unknown participants and still be very understandable, preserving its value.

XML has both the descriptors of the data and the data in the same transaction or data record, and this single characteristic has made it far more flexible and enabled more widespread data exchange than any previous data communications method.

Chapter Five

Here's a more detailed and formal definition of XML, from Wikipedia (the bolded italics are mine):

> XML (Extensible Markup Language) is a general-purpose specification for creating custom markup languages. It is classified as an extensible language, because it allows the user to define the mark-up elements. XML's purpose is to aid information systems in ***sharing structured data***, especially via the Internet; to encode documents; and to serialize data; in the last context, it compares with text-based serialization languages such as JSON, YAML and S-Expressions.
>
> XML's set of tools helps developers in creating web pages but its usefulness goes well beyond that. XML, in combination with other standards, makes it possible to define the content of a document separately from its formatting, making it easy to reuse that content in other applications or for other presentation environments. ***Most importantly, XML provides a basic syntax that can be used to share information between different kinds of computers, different applications, and different organizations, without needing to pass through many layers of conversion.***
>
> XML began as a simplified subset of the Standard Generalized Markup Language (SGML), meant to be readable by people via semantic constraints; application languages can be implemented in XML. These include XHTML, RSS, MathML, GraphML, Scalable Vector Graphics, MusicXML, and others. Moreover, XML is sometimes used as the specification language for such application languages.
>
> XML is recommended by the World Wide Web Consortium (W3C). It is a **fee-free open standard.** The recommendation specifies lexical grammar and parsing requirements.

XML is still in its infancy in relative terms. In ten short years, it has spawned many industry efforts to standardize data, and we'll take a tour of these efforts later in the chapter. My heartfelt prediction is that XML's best years are still to come, and its impact on the world will be significant.

Key benefits of XML

It's free. All the organizations and businesses involved with creating XML agreed to make it a free and open standard. Many companies could have stood in the way of this through variously held patents and other intellectual property, but the W3C (World Wide Web Consortium – a group that did important work to bring the web to the world) pushed for a freely available standard and ultimately prevailed. In 100 years, this strategy and approach will be looked back on as an effort that greatly benefited civilization.

It's extensible. XML is a very strong markup and descriptive standard that's extremely flexible, based on the application. Using the basic tenets of the markup language, widely varying industry-specific vocabularies can be created to serve any need, from covering the range of Beanie Baby categorization to the exchange of highly structured genetic model transfer information. This is a relatively simple concept of immense importance. It is similar to using the 26 letters of the English alphabet to create words, using various rules for pronunciation and construct, and enforcing a process for defining what new words mean, thus creating a complicated language capable of conveying an immense range of simple and complex ideas…using only 26 letters.

It can be automated. XML allows machines to easily exchange information. It has a very structured syntax so that it can be automatically checked for errors – ensuring that the data exchanged is "well formed" and syntactically correct. Moving structured and intelligent data seamlessly between machines will enable a whole new set of applications – imagine if your dentist could query your calendar and suggest appointments based on an intelligent assessment of options.

What does XML look like?

In this very simple example, the XML describes a note that can be moved from one system to another. As you can see, much of the information is readable by humans, but in a structure that a machine can evaluate and understand.

```
<?xml version="1.0" encoding="ISO-8859-1" ?>
    <note>
        <to>Pete</to>
        <from>Sherry</from>
        <heading>Reminder to get this book published</heading>
            <body>Don't forget to review the changes I made!</body>
    </note>
```

A brief history of XML

XML grew out of SGML (Standard Generalized Markup Language), a very complex standard used by digital media publishers even before the Internet's meteoric rise. The W3C added the XML effort to its growing responsibilities in 1995, and work began in 1996. Eleven member working groups and over a hundred interest groups worked together, mostly via teleconferences and email. XML 1.0 was recommended by the W3C in February of 1998. Relevant goals for XML were:

- Internet usability: The wildly popular Internet demanded methods more sophisticated than HTML.

- General purpose stability: XML needed to be usable to solve a wide range of problems rather than be highly specialized to an industry or a technology. You will see by the wide variety of use that this goal was exceeded.

- Formality: A precise structure allows for broad adoption and a higher assurance that if you follow the rules and syntax, your efforts will work anywhere.

- Concision: Precise and consistent language and punctuation without a verbose, complicated structure.

- Ease of authoring: Any text editor or word processor (or multiple other devices) can create a valid XML document.

- Minimal new features: Rapid change doesn't bode well for massive adoption or a uniform deployment.

The people that birthed the XML specification deserve tremendous credit for their efforts – this was truly a world-changing effort.

Technical standards using XML

There are a large number of development efforts under way that use XML to define standards for future web activities – authentication, publishing, and many others. These standards are very important to the future of the web, but they're too complex for an overview here to be fair treatment. Two very good places to look for more in-depth information are www.W3C.org and www.oasis.org. Both these organizations are sophisticated and effective; I recommend that someone in your organization join them as your involvement in standardized data grows.

Chapter Five

Major industry standardized efforts using XML

Here is a short tour covering many of the non-technical industry efforts to create a common XML-based vocabulary for specified purposes and industries. In the next chapter, we'll dissect a couple of easy-to-understand standards so that the actual meaning becomes clear at the data field level. You will see that there is tremendous interest and momentum for establishing these vocabularies as building blocks for future initiatives, and for smoothing the way for faster transactions and commerce within industries. As various industries interlock into a distribution vocabulary, you can see how the seamless flow of information will be accomplished over time.

1. **Astronomy.** NASA's FITSML (Flexible Image Transport System Markup Language) is a data format designed to provide a means for convenient exchange of astronomical data between installations whose standard internal formats and hardware differ. The "Image'" in FITSML comes from the original use of the format to transport digital images, but it's not just for images anymore. See http://fits.gsfc.nasa.gov.

2. **Built environment, and infrastructure systems integration.** The oBIX (Open Building Information Xchange) is a focused effort by industry leaders and associations working toward creating standard XML and Web Services guidelines to facilitate the exchange of information between intelligent buildings, enable enterprise application integration, and bring forth true systems integration. Based on standards widely used by the IT industry, the oBIX guidelines will improve operational effectiveness, giving facility managers and building owners increased knowledge and control of their properties. Comprised of representatives from the entire

Silver Bullets

spectrum of the buildings systems industry, oBIX includes professionals from the security, HVAC, building automation, open protocol and IT disciplines. See www.obix.org.

3. **Distribution/Commerce.** The RosettaNet consortium is a global forum for suppliers, customers, and competitors to come together and create industry-specific standards so collaborative international commerce can proceed in an efficient and profitable manner. RosettaNet standards enable companies to meet the various legislative demands of their countries, extend their collaborative networks, and achieve user-specified goals. Based on a track record of proven business value and measurable results, RosettaNet standards are spreading throughout the globe with user-defined best practices to achieve efficient, collaborative commerce. See www.rosettanet.org.

4. **Education.** SIF (Schools Interoperability Framework in the U.S., Systems Interoperability Framework in the U.K.) is an open data-sharing specification for academic institutions from kindergarten through twelfth grade (K-12). Until recently, it has been used primarily in the United States; however, it is increasingly being implemented in Australia, the U.K., India, and elsewhere. The specification is composed of two parts: an XML specification for modeling educational data, and a Service-Oriented Architecture (SOA) specification for sharing that data between institutions. The SIF Association brings together the developers and vendors of school technologies with the federal, state, and local educators who use those technologies, to define the rules for data movement between applications for efficiency, accuracy, and automation. See www.sifinfo.org.

Chapter Five

5. **Financial reporting.** XBRL (eXtensible Business Reporting Language) is a language for the electronic communication of business and financial data that is revolutionizing business reporting around the world. It provides major benefits in the preparation, analysis, and communication of business information. It offers cost savings, greater efficiency, and improved accuracy and reliability to all those involved in supplying or using financial data. XBRL is being developed by an international non-profit consortium of approximately 450 major companies, organizations, and government agencies. It's an open standard, free of license fees. It is already being put to use in a number of countries, with global implementations growing rapidly. The U.S. SEC is mandating XBRL reporting for major companies as a means of bringing transparency and uniformity to quarterly and annual reports. See www.xbrl.org.

6. **Financial research.** RIXML is a consortium of buy-side firms, sell-side firms, and vendors that have joined together to define an XML-based open standard for categorizing, tagging, and distributing global investment research. The RIXML standard provides extensive capabilities to tag any piece of research content, in any form or media, with enough detail for end users to be able to quickly search, sort, and filter aggregated research. RIXML is creating an open specification that can be freely used by application vendors, research providers, and their clients. See www.rixml.org.

7. **Food.** The Meat and Poultry B2B Data Standards Organization (mpXML) is pioneering the development and use of standards to support ecommerce across all segments of the meat and poultry supply chain. Trading partners aim to develop interoperable standards for voluntary adoption throughout

the supply chain, and will be able to assist actively and comment on electronic messaging and product identification systems as they are developed. See www.mpxml.org.

8. **Healthcare.** Health Level Seven (HL7) is a not-for-profit Standards Development Organization (SDO) dedicated to developing and providing a comprehensive framework and standards for the exchange, integration, sharing, and retrieval of electronic health information to support clinical practice and the management, delivery, and evaluation of health services. HL7 provides standards for interoperability that improve care delivery, optimize workflow, reduce ambiguity, and enhance knowledge transfer among stakeholders including healthcare providers, government agencies, the vendor community, fellow SDOs, and patients. HL7's 2,300-plus membership includes approximately 500 corporate members representing over 90 percent of the information systems vendors serving healthcare. See www.hl7.org.

9. **Information technology architecture.** ADML (Architecture Description Markup Language) is being developed as a standard for communicating the detailed aspects of IT architecture between architecture tools, and over a system's lifecycle. ADML Version 1 is the initial version of the ADML standard, based on technology developed by the Microelectronics and Computer Technology Consortium (MCC). See www.opengroup.org.

10. **Instruments.** Instrument Markup Language (IML) is an XML specification that applies to virtually any kind of instrument that can be controlled by a computer. The approach to instrument description and control applies to many domains,

from medical instruments to printing presses to machine assembly lines. The concepts behind IML apply equally well to the description and control of instruments in general. IML is a co-project of NASA's Goddard Space Flight Center and Commerce One. See www.nasa.gov

11. **Insurance.** ACORD (Association for Cooperative Operations Research and Development) is a global, nonprofit standards development organization serving the insurance industry and related financial services industries. ACORD's mission is to facilitate the development of open consensus data standards and standard forms. ACORD members include hundreds of insurance and reinsurance companies, agents and brokers, software providers, and industry associations worldwide. ACORD works with these organizations toward a goal of improved data communication across diverse platforms through implementation of standards. See www. acord.org.

12. **Legal.** LegalXML is a member section within OASIS (Organization for the Advancement of Structured Information Standards), the not-for-profit, global consortium driving the development, convergence, and adoption of ebusiness standards. LegalXML brings legal and technical experts together to create standards for electronic exchange of legal data. Members set the LegalXML agenda, using the open OASIS technical process expressly designed to promote industry consensus and unite disparate efforts. LegalXML produces standards for electronic court filing, court documents, legal citations, transcripts, criminal justice intelligence systems, and others. OASIS members participating in LegalXML include lawyers, developers, application vendors, government agencies, and members of academia. See www.legalxml.org.

13. **Manufacturing.** The PSLX (Planning and Scheduling on XML Language) consortium is an international group working to establish an APS (Advanced Planning and Scheduling) standard for collaborative manufacturing, and to support the implantation of the standard by manufacturers world-wide. See www.pslx.org.

14. **News.** The NITF (News Industry Text Format) uses XML to define the content and structure of news articles. It's being driven by the International Press Telecommunications Council (IPTC), a consortium of the world's major news agencies, news publishers, and news industry vendors. The IPTC develops and maintains technical standards for improved news exchange (including content, metadata and management metadata) that are used by virtually every major news organization in the world. About 70 companies and organizations from the news industry are members, drawn from all continents except South America. See www.iptc.org.

15. **Oil and gas.** The Petroleum Industry Data Exchange (PIDX) is the American Petroleum Institute's (API) committee on Electronic Business Standards and Processes, sponsored by its General Committee on Information Management & Technology (GCIMT). PIDX is dedicated to helping the industry and individual companies improve the efficiency, effectiveness, and value of the information management, business process, and technology functions within the oil and natural gas industry and its trading partners. See www.pidx.org.

16. **Publishing.** DocBook is a schema (available in several languages including RELAX NG, SGML and XML DTDs, and W3C XML Schema) maintained by the DocBook Technical

Committee of OASIS. It is particularly well suited to books and papers about computer hardware and software (though it is by no means limited to these applications). Because it is a large and robust schema, and because its main structures correspond to the general notion of what constitutes a "book," DocBook has been adopted by a large and growing community of authors writing books of all kinds. DocBook is supported out of the box by a number of commercial tools, and there is rapidly expanding support for it in a number of free software environments. These features have combined to make DocBook a generally easy to understand, widely useful, and very popular schema. Dozens of organizations are using DocBook for millions of pages of documentation, in various print and online formats, worldwide. See www.oasis.org.

17. **Real Estate.** The Real Estate Transaction Standard (RETS) facilitates data transfer between partners in the real estate industry. Creating and improving RETS is a collaborative effort to simplify moving real estate information from system to system, as well as simplify solution development efforts. As RETS usage matures and expands, MLS (Multiple Listing Services) with geographic overlaps can create data-sharing policies that provide their members a single point of entry to search multiple MLS data sets. See www.RETS.org.

18. **Research.** The Consortia Advancing Standards in Research Administration Information (CASRAI) is a not-for-profit organization which addresses a growing problem: Conducting and administering research today is an increasingly multi-stakeholder and multi-disciplinary endeavor. The stakeholders include the highly qualified personnel doing the actual research (researchers and students) and the organizations

that facilitate and support the research (universities, colleges and funding agencies). This ecosystem of independent but collaborative stakeholders depends on an increasingly fragmented, duplicative, and complex set of data about research personnel and activities. Recording, maintaining, analyzing, and sharing this data is difficult, and places a heavy administrative burden on researchers. CASRAI aims to solve this problem by standardizing the information that must be collected and shared so that a single, authoritative source of data can serve the needs of all stakeholders (write-once, reuse-anywhere). Once the semantics and format of the source data are standardized, it can be reliably collected, maintained and shared using any software hosted at any location. See www.casrai.org.

19. **Telecommunications.** The Alliance for Telecommunications Industry Solutions (ATIS) prioritizes the industry's most pressing technical and operational issues, and creates interoperable, implementable, end-to-end solutions – standards when and where the industry needs them. Over 600 industry professionals from more than 250 communications companies actively participate in ATIS committees and incubator solutions programs. ATIS develops standards and solutions addressing a wide range of industry issues in a manner that allocates and coordinates industry resources and produces the greatest return for communications companies. See www.atis.org.

20. **Travel.** The OpenTravel Alliance (OTA) is a member-funded, nonprofit organization formed in 1999 by major airlines, hoteliers, car rental companies, and companies that provide distribution and technology systems to the

Chapter Five

travel industry. OpenTravel's primary activity is to develop and maintain a library of XML schemas for use by the travel industry, enabling suppliers and distributors to speak the same interoperability language. These schemas constitute the OpenTravel XML specification, which is based on the W3C XML Schema standard. See www.opentravel.org.

Challenges using XML-based data

As with all young technologies, XML has limitations that call for improvement or require compensatory technologies. None of these should dissuade you from pressing forward, and there are multiple methods available to build in further security, error checking, and semantics. These are often outside the formal XML specifications, added as supplementary capabilities. Areas to note include:

1. **Source verification at the file or transaction level.** XML cannot tell you whether a file came from a specific sender. Source verification is important, particularly if there's risk of someone sending erroneous or falsified information. This can be particularly important in real-time situations such as tsunami warnings or severe weather alerts. Technologies such as encryption, point-to-point communication links, and SSL can all be used to mitigate this risk.

2. **Field verification beyond a value range.** XML can enumerate a range of values for a field, but has no capability to do a real-time lookup on a value to ensure that it is valid and based on some centralized or distributed data source. This isn't necessary in many cases, but there are dynamic situations where having this type of functionality over a wide range of organizations would be very helpful. For example, the XML-based public safety standard format Common Alerting

Protocol (CAP) has a unique identifier field, but no inherent way to enforce this uniqueness across CAP issuers. Having a dynamic lookup capability to ensure validity, and the ability to obtain additional information about the originator would be desirable, but would require a DNS- (domain name service) like capability, which is a distributed method for validating internet URLs around the world.

3. **Cross-industry and cross-domain communications and reconciliation.** A common ground among the many sector-specific XML vocabularies would be helpful. For example, if a "citizen" in a criminal vocabulary is defined differently than in a voting vocabulary, you can guarantee confusion if the two types of information need to be mixed for future analysis. The National Information Exchange Model (NIEM) has been working to reconcile vocabulary across the domains it supports, and you'll see this develop over time.

4. **Verbose and partially redundant.** XML has a substantial number of special characters that define its structure, and this may make the information harder for humans to read and process if it has to be created or read manually. Fortunately, most of the information will be handled by a series of application programs or utilities designed to ensure that the information is well formatted and valid.

5. **Non-hierarchical data can be challenging.** XML is aimed at hierarchical data structures, which address a variety of situations (a fire station has one commander, four trucks, 22 firemen, one Dalmatian.), but there are other models where the logic may outstrip XML's ability to organize this information easily.

Chapter Five summary

1. XML is a rapidly developing, global method for data exchange and transmission between organizations.

2. XML is very flexible and can handle many different circumstances, and is also rigorous in the formatting of data, so that machines can ensure accuracy of the contents.

3. There are many industry-specific efforts to define XML vocabularies. Your organization should leverage this work if possible – far better to take advantage of other work than to try to recreate the effort.

4. XML is critical for cross-organizational information sharing; partners can build exchanges very quickly if the transaction format is defined and everyone agrees on usage.

5. There are many standards bodies working hard to define functional XML vocabularies and move them into the mainstream of industry, government, and other organizations. These cross-industry efforts, done well, will pay large dividends for civilization.

6. The power of XML transactions is in their uniformity and usage – millions of data transfers are taking place today, and the trend is continually upward.

7. Machines can generate XML themselves. Sensors may sit idle for months, then send an XML-structured alert to a central collection site when some parameter (example: an air quality benchmark) is exceeded. By the same token, other sensors might generate 30 transactions per minute, but only deviations from a normal set of parameters would generate further processing.

8. If your organization isn't using XML to strategic advantage, you should find out why. This is not a technical discussion, but one that impacts the strategic posture of most organizations.

Chapter Six

Two great interoperable data standards: CAP and KML

> *Customers need to be given control of their own data – not tied into a certain manufacturer so that when there are problems they are always obliged to go back to them.*
> —Tim Berners-Lee

Two important interoperable data standards have begun to move into global prominence over the last several years. Each is excellent in terms of structure and function, and both are reasonably simple and clear examples of interoperable data structures. The two standards are nicely complementary to each other, but come from vastly different origins. The Common Alerting Protocol (CAP) has come from a public/private organization, the Partnership for Public Warning (PPW). Keyhole Markup Language (KML) was created by a for-profit company, Keyhole, which has been acquired by Google.

Both standards use XML for their underlying format and structure. Each is enjoying rapid growth in its global base of adopters, and each has an active set of consensus-based discussions underway for future improvements on a worldwide basis.

Common Alerting Protocol (CAP) – alerts and warnings

Imagine you're hiking in a southwestern canyon with your children. The sky is blue, the temperature comfortable. You're enjoying the beauty, but you're also aware that this area can experience flash floods. Your cell phone receives a text message warning of a cloudburst upstream that means you and your children could be in danger. You scramble out of the canyon to higher ground – and fifteen minutes later, a torrent of water comes rushing through. The text message warning you received is an example of the power of CAP, a standardized format that enables a broad range of organizations to send alerts and warnings to a broad range of receiving devices. It's a standard you'll be betting your life on, as it's implemented worldwide.

CAP is a noteworthy example of how a standardized interoperable data format can be used across many computer systems and devices, compounding its value and enabling far-reaching capabilities that can cross continents, save lives, and prevent or minimize property damage – all at the speed of the Internet and other modern communication methods.

As its name implies, CAP was designed as a standard method of creating and distributing alert and warning messages across a broad range of cultural and technological variables (devices, networks, languages, mapping formats, and much more). Alerts and warnings are especially time critical – it's of little use to have a tsunami or tornado alert arrive two hours after the affected area has been destroyed – talk about rubbing salt in the wound!

Alerting and warning is a difficult problem to solve despite thousands of years of trial and error. We've only recently become able to address this problem on a global basis with the revolution in digital communications. If you are in a building and see a fire, you can yell "Run!" or hit the fire alarm. Larger alerts demand a much broader solution, and the size can impact the difficulty. See "Strawberry Shortcake for 400,000" as a graphic example of how scale affects complexity:

> **How something that appears simple can quickly become complicated:**
> *Strawberry Shortcake for 400,000*
>
> Roger Von Oech, in his book *A Kick in the Seat of the Pants*, gives an example of how complexity grows as a problem increases in size and scope. He starts out with a recipe for fresh strawberry shortcake for four people, which anyone could make in today's world – just zip over to Whole Foods, and voila! He then asks how the problem would change if you suddenly had to produce 400,000 servings of strawberry shortcake and address the logistical and operational issues. For example:
>
> - How much lead time do we need to get this done?
> - Do we need permits from the City?
> - How many tankers of cream do we need? Where can we park them?
> - How do we whip the cream into something we can serve?
> - Where can we source enough strawberries?
> - Where will all the people sit? Where will they park their cars?
> - What happens if it rains?
> - Where does the trash go?
>
> The story illustrates that in order for a standard to scale to global capability, it must be able to address the exponentially growing issues of usability, completeness, flexibility, and scalability.
>
> I think it also demonstrates why you haven't been invited to a dessert party with 399,999 other people – the risk/reward quotient of this endeavor makes it not worth doing.

The creators of CAP developed an excellent design with the initial standard and decided to work with the Organization for the Advancement of Structured Information Standards (OASIS) in order to further the standard and help develop a good strategy to improve CAP over time. OASIS was chosen for three major reasons:

- OASIS is an international standards organization, and CAP addresses a worldwide alerting and warning problem.
- OASIS provides a consensus-based process that's open to anyone in the world who wants to comment or provide input, and all inputs are visible to the general public.
- The OASIS standards are free to download and use, which removes a cost barrier to adoption.

If big projects like a global implementation of CAP were truly easy, they likely would have been accomplished already. While many governments are deemed guilty of accomplishing little or nothing in terms of innovation (perhaps in part because of the seemingly overwhelming issues), it is a reality that scaling a solution to a large deployment can be difficult, time consuming, expensive, and has more than a chance of total failure.

History of CAP

The U.S. National Science and Technology Council report "Effective Disaster Warnings," released in November 2000, made this recommendation: "A standard method should be developed to collect and relay instantaneously and automatically all types of hazard warnings and reports locally, regionally and nationally for input into a wide variety of dissemination systems."

CAP was formally conceived and begun by the Partnership for Public Warning in November 2001 – at a meeting that had been scheduled before the 9/11 attacks, but took on special significance in light of these tragedies. 130 people attended, including Art Botterell and Elysa Jones, who were to become key drivers of the specification.

A draft specification was created, and several ground-breaking demonstrations were executed over subsequent months and years. The Emergency Interoperability Council, another public/private consortium, assisted in the development and evangelizing of CAP, and OASIS published the CAP 1.0 Specification in April 2004. Minor adjustments were made, and CAP 1.1 was approved in October 2005. An erratum was issued in 2007, and the International Telecommunications Union endorsed CAP.

Many organizations adopted the standard and implemented it in the early adopter phase. RAINS (Regional Alliance for Infrastructure and Network Security) in Portland, Oregon, adapted the local 9-1-1 center's dispatch call data into a CAP format, and created a program named Connect & Protect to supply alerts to trusted and vetted community members in near-real time. The program linked a broad spectrum of public safety stakeholders, including mall security, business owners, hotel security staff, school principals, and local government officials. The 9-1-1 information was blended with weather information, aggregated traffic data, and other information feeds – all integrated into a map/common operating picture mash-up. The user interface displayed the map with points of interest representing incidents, together with a rotating set of video camera feeds from around the city, and an updated RSS feed providing web-published information about Portland. (See "Reinventing 911" by Gary Wolf, *Wired* Magazine, December 2005.)

Because the CAP standard was in the public domain, there was no coordination necessary between RAINS and the CAP working group – further showing the power of an open published standard.

Information about incorporating CAP-formatted alerts was usually available from the source. For example, someone planning to use CAP to create a warning service providing alerts from the National Oceanic and Atmospheric Administration (NOAA) could find all the necessary information freely available on the web. Both the OASIS website and www.incident.com include specifications from NOAA for integration of their severe weather alerts, information on the NOAA earthquake and tsunami alerts, and more.

CAP in 2010

CAP adoption has grown rapidly in the last several years, and one of the classic web models is beginning to be realized – the network effect, in which adoption spurs value in an upward growth spiral. When this spiral occurs, more information providers utilizing CAP cause more software and technology providers to adopt CAP as a format, which generates more utilization of CAP by the consumers of the standardized alerts and warnings, causing publishers to adopt the format for export, and so on in a broadening circle of wider adoption. These end users reinforce the method as a standard, pushing more information providers to utilize the published CAP format, and the upward spiral increases even more. With producers, distributors, and consumers increasing adoption, a standard can advance very quickly and gain critical mass.

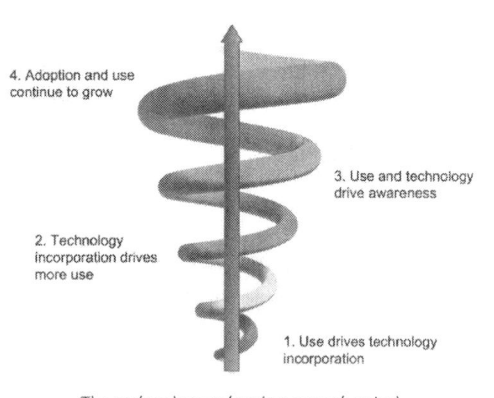

The technology adoption growth spiral

1. Use drives technology incorporation
2. Technology incorporation drives more use
3. Use and technology drive awareness
4. Adoption and use continue to grow

Chapter Six

Why CAP is succeeding

CAP is a growing standard for many different reasons, but the keys have been its tenacious creators/advocates and the CAP core design tenets agreed upon early in the process. CAP is…

- **Simple.** CAP is a straightforward, linear message format that non-programmers and technical personnel alike can relate to once the inherent elements are described. It consists of just three major elements, which keeps things simple.

- **Interoperable.** CAP is designed to work across many systems, with current and legacy devices, and to remain usable for future devices as they evolve. CAP is XML- based, which makes it consumable, able to be reformatted, and adaptable to both simple and complex situations.

- **Complete.** CAP has all the elements necessary to issue, update, withdraw, and expire alerts and warnings, so it provides a complete closed-loop information sharing environment in one convenient package.

- **Multi-use.** CAP alerts and warnings can be adapted to many different types of situations and alerting situations:

    ▸ Local alerting: 9-1-1 alerts for a local municipality can be distributed in near-real time to the public or to preselected members of that community. A school system could deploy a system to warn all its schools, while a hospital could send alerts to selected emergency room physicians.

    ▸ Regional alerting: Multiple groups can team up and send cross-jurisdictional information not normally available in near-real time. Imagine one county getting an alert of a high-speed chase heading across county lines. The first responders on the radio would know, but a CAP alert could inform other impacted agencies, trigger action by support personnel such as changes to electronic signage, and potentially update GPS-based traffic systems with a warning. Regional intelligence centers could alert each other about events as they occur, looking for verification of concurrent incidents of the same type.

- Global alerting: In 2004, an undersea earthquake in the Indian Ocean triggered a large-scale tsunami in Indonesia. Both the Alaska and Pacific (Hawaii) NOAA tsunami warning centers knew that it happened, and that a major wave would occur. They generated warnings, but there was no way to reach the thousands of local communities that needed to know. Close to 300,000 people on both sides of the ocean perished. A simple alert, delivered to PCs, cell phones, and other devices in near-real time (and in the right language), merely advising people to move 200 yards inland, could have saved countless lives. With CAP, alerts generated anywhere in the world can be delivered globally, in near-real time, to a wide range of stakeholders: public safety officials, hotels, transportation providers and many more.

- Cross-organizational: CAP alerts can be exchanged between organizations on either a public or private basis. This is not a function of the standard, but of the policy and infrastructure agreed upon by the parties sending and receiving. Cross-organizational information sharing is critical for government as it tries to communicate with a broad group of stakeholders. Imagine the alerting that would need to take place if there was a known threat against a major Fortune 500 company, but the government didn't know which one. In a large-scale emergency, there can be hundreds of public and private organizations participating, culminating in the kind of confusion seen in the aftermath of Hurricane Katrina.

  ▶ Intra-organizational: CAP can be used inside large or multi-location corporations without being shared externally, allowing these organizations a way to share alerts and warnings confidentially. When a situation arises mandating that information must be disclosed, a common format is already in place. Many global corporations have thousands of locations with people and assets around the world. Having a fast, reliable means to share important information can be critical, yet right now it's often done with relatively primitive means.

- **Familiar.** The elements of a CAP alert are highly familiar and were drawn from the collective emergency and incident experience of the groups that devised the standard. "Who," "what," "when," "where," "why," and "how" are good starting points for almost any exploration into a standard transaction.

- **Interdisciplinary.** CAP alerts and warnings are not limited to emergency groups (such as first responders), or geophysical events (such as weather or earthquakes). The format was designed to handle a wide range of alerting and warning situations.

Silver Bullets

- **International.** CAP was designed to support multiple languages within the same alert context, making it ideal for use worldwide or when alerts must be delivered in multiple languages within a geographic area.

CAP examples. The images below illustrate various ways CAP might be displayed to a recipient.

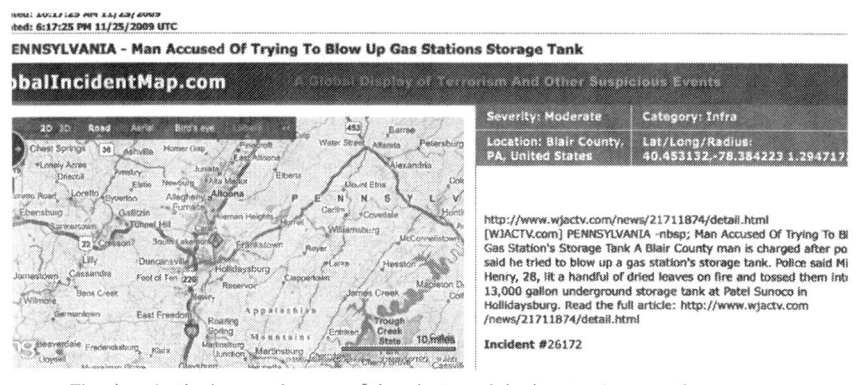

The data in the boxes, the text of the alert, and the locator icon on the map are pulled from the data in a single CAP alert and displayed on a (full color) user interface

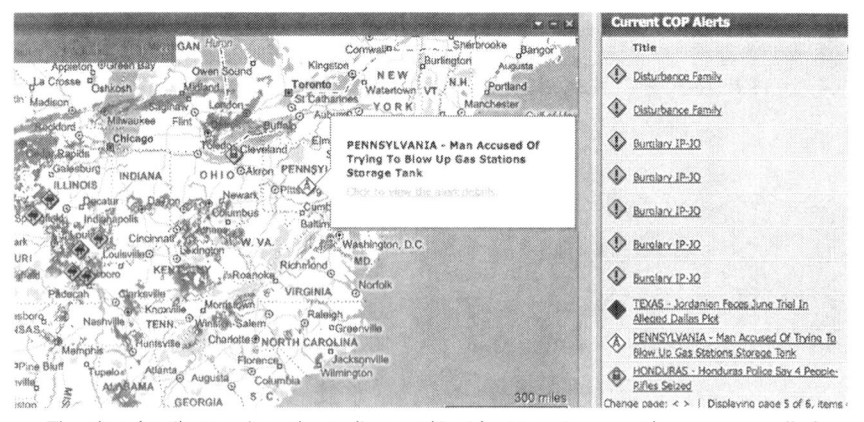

The alert details, severity color coding, and incident type icons on the map are pulled from the data in multiple CAP alerts and displayed on a (full color) user interface

100

Chapter Six

What the XML coding of a CAP alert looks like:

```
<alert xmlns="http://www.incident.com/cap/1.0">
    <identifier>26172</identifier>
    <sender>http://www.wjactv.com/news/21711874/detail.html</sender>
    <sent>2009-11-25T10:17:25.2092277-06:00</sent>
    <status>Actual</status>
    <msgType>Alert</msgType>
    <info>
        <category>Infra</category>
        <event>Oil Gas Infrastructure - Incidents / Threats/ News</event>
        <urgency>Past</urgency>
        <severity>Moderate</severity>
        <certainty>Unlikely</certainty>
        <headline>PENNSYLVANIA - Man Accused Of Trying To Blow Up Gas Stations Storage Tank</headline>
        <description>[WJACTV.com] PENNSYLVANIA - Man Accused Of Trying To Blow Up Gas Station's Storage Tank A Blair County man is charged after police said he tried to blow up a gas station's storage tank. Police said Michael Henry, 28, lit a handful of dried leaves on fire and tossed them into a 13,000 gallon underground storage tank at Patel Sunoco in Hollidaysburg.   Read the full article: http://www.wjactv.com/news/21711874/detail.html</description>
        <area>
            <circle>40.453132,-78.384223 1.294717248</circle>
        </area>
    </info>
</alert>
```

The future of CAP

CAP is being driven by a comprehensive committee inside OASIS, chaired by Elysa Jones, Chief Technology Officer of Warning Systems. OASIS recognized Jones with an award for her long and inspired service to make CAP a successful standard.

CAP 1.2 will contain changes that will allow the Federal Emergency Management Agency (FEMA) adopt CAP across several U.S. national alert and warning systems. CAP 2.0 will be another step forward in terms of the functionality and capability needed for global alerting systems that can interoperate, cooperate, and push information to a wide variety of targeted recipients. Driven by the standards process, it will allow for consensus and interlock with other emerging standards in the emergency management and public safety realm. It is also intended to be compatible with GML (Geospatial Markup Language) and more fully support the EDXL/DE (Emergency Data Exchange Language Distribution Element).

CAP is intersecting other interoperable work being done in the Emergency Management Domain of the National Information Exchange Model and OASIS. RM (Resource Messaging) and HAVE (Hospital Availability Exchange) are new standards that will help manage delivery and routing; resource allocation, sourcing, loans and returns; and the exchange of hospital information to people who need it.

CAP adoption and usage

As described previously, CAP is experiencing an upward growth spiral as more awareness, adoption, technological incorporation by vendors, and the increasing volume of alerts work together to drive the standard into worldwide use. The World Meteorological Organization (WMO) has adopted the format for global weather alerts, and far-flung countries such as Sri Lanka are evaluating CAP.

My prediction is: You will see CAP in every type of alerting system across a broad range of organizations including governments, NGOs, and corporations. As interoperability grows and sharing occurs, all concerned will have a much better picture of the many threats around them, and will be able to take proactive action where possible. All this is occurring because some very committed and smart people had a vision for a unified and interoperable method for alerts and warning.

Chapter Six

KML – Keyhole Markup Language (Open Geospatial Consortium and Google)

With Google Earth and Google Maps in the last several years, Google has become a powerhouse in the consumer mapping and visualization industry. CNN uses Google Earth heavily in its visual presentations of events and incidents on its 24-hour TV news network, using a special touch interface to easily zoom and pan to different areas. Many of the "mash-ups" – maps with data sets layered upon them – are developed using Google's underlying mapping engine.

A small company named Keyhole Corporation created a very exciting product called EarthViewer 3D. One of the key features in this geospatial suite of products was Keyhole Markup Language (KML). Keyhole Corporation was acquired by Google in 2004, and the Google/Keyhole team has subsequently improved the product greatly. Usage exploded with the relative ease of use that these geospatial products brought to the marketplace.

> KML is an XML language focused on geographic visualization, including annotation of maps and images. Geographic visualization includes not only the presentation of graphical data on the globe, but also the control of the user's navigation in the sense of where to go and where to look.
>
> —*The Open Geospatial Consortium*

If you think of Google Earth and Google Maps as geospatial-browsers ("geobrowser" for short—a web interface designed to show spatially oriented information), you will see that KML is the HTML equivalent for the geobrowser. KML drives what is being displayed on the map or globe.

Silver Bullets

Rather than lock customers into a proprietary format or means of interfacing their data, Google submitted KML to the Open Geospatial Consortium (OGC) for inclusion as a standard in 2007. The OGC is a non-profit, international standards organization committed to geospatial and location-based standards worldwide. KML 2.2 was officially announced as an OGC standard on April 14, 2008.

The OGC has worked to ensure that KML is complementary to most other key existing OGC standards including GML (Geography Markup language), WFS (Web Feature Service), and WMS (Web Map Service).

KML was not the first mapping language to be developed; there were already multiple formats for moving data from one vendor to another. KML, once adopted by the OGC, was then free, open, and less affected by the vendor community, so it has quickly become the standard of choice.

Like CAP, KML is designed with inherent flexibility and can handle the very simple (show my three schools on a map) to the incredibly complex (show me a visualization of a volcanic eruption over a period of six weeks). KML is much more complex than CAP, and requires more of its users, including in-depth knowledge of coordinate systems, cartography, and other geospatial concepts used to work with complex models. Best practices have emerged for KML users around the world, with tremendous sharing and collaboration around the many varied uses of KML.

One key feature of KML is the ability to compress a large file of geographic information into a much smaller archive file (a KMZ file) for transport and movement between users. KMZ uses the ZIP file format to greatly enhance portability around the web.

Chapter Six

KML is also very flexible for organizations, because it can be extended by anyone or any organization, simply by appending a new namespace to the XML block. Google and others can add features that are not supported in the current standard, but will be considered for adoption in later versions of the standard. This approach makes the standards-setting process much more dynamic and practitioner-based than it would be if achieved solely through academic discussions. Adoption can be looked at based on live feedback and usage patterns that have already occurred.

Here are examples of KML usage (source: OGC KML 2.2 Specification):

- Annotate the Earth
- Specify icons and labels to identify locations on the surface of the planet
- Create different camera positions to define unique views for KML features
- Define image overlays to attach to the ground or screen
- Define styles to specify KML feature appearance
- Write HTML descriptions of KML features, including hyperlinks and embedded images
- Organize KML features into hierarchies
- Locate and update retrieved KML documents from local or remote network locations
- Define the location and orientation of textured 3D objects

KML implications and future

You are seeing an important world standard emerge before your very eyes. In a few short years, Google and the OGC have positioned KML to be one of the primary global standards for geospatial capabilities and also ensured interoperability between the multitude of vendors that make up the diverse community surrounding geospatial representation.

Google could have tried to maintain KML themselves, charge for every usage and generally attempt to lock users into their products. Google's open approach of putting KML into the OGC standards group has helped drive the capability to many more uses and across a wider range of global users. The KML standard is supported by virtually all of the major proprietary GIS vendors, including ESRI, Intergraph, Autodesk and others – moving the entire industry toward more interoperability.

It is hard for most proprietary vendors to understand, but everyone wins when open standards help drive markets to be much bigger. While stories abound of companies that have struggled or sunk by holding onto proprietary approaches (Wang, Digital Equipment Corporation, Silicon Graphics, and Encyclopedia Britannica are a few that come to mind), there are few examples of companies embracing an open standard and being punished for it.

Chapter Six

Chapter Six summary

1. Common Alerting Protocol (CAP) is an easy-to-understand, well-developed standard emerging worldwide as an information-sharing format for alerts and warnings.

2. Keyhole Markup Language (KML) gives the average user an incredible amount of power and flexibility for creating custom map mash-ups. Version 2.2 has been submitted and accepted as a standard by the OGC (Open Geospatial Consortium), a non-profit organization committed to making information sharing across geospatial environments possible.

3. Both of these standards are relatively new, and have come from very different pedigrees. CAP emerged from a group of concerned individuals responding to a government call for action. KML came as a gift from a commercial geospatial effort to a worldwide standards organization.

4. Adoption and use are key to an interoperable data format's success. If a community at large adopts the format and reinforces its use, growth can occur quickly, and the standard can last a very long time. Both CAP and KML are gaining substantially larger amounts of use each year.

5. While CAP's technical aspects are interesting to developers, for most people CAP's promise is to deliver the value of adopting standard formats and integrating them worldwide.

6. Both standards will continue to be improved, and begin to integrate with other efforts in the emergency management and public safety worlds.

7. Rapid adoption of these standards could help push other standards as people realize the time and cost savings of using interoperable data transactions.

Case Study

Golden Phoenix

In July 2008, a large training event was held in San Diego (in several military, civilian, and educational locations), and Yuma, Arizona. It was conceived and developed by LTC John Persano with Marine Aircraft Group-46 (Marine Forces Reserve) and in 2008, co-led by George Bressler of the U.S. Customs and Border Protection. It was supported in a number of ways by the Department of Homeland Security, San Diego State University, and the DoD.

More than 150 organizations including local, state, federal, and tribal agencies and academic, nongovernmental, and private sector entities participated, for a total of over 700 people. Prior to the three-day live event, there was an intelligence briefing for both the classified and unclassified groups, giving them background information on the scenario, and allowing discussion and brainstorming around the elements of the event.

The scenario was simple but devastating. Ten people were caught coming across the U.S.-Mexico Border with what looked like methamphetamine – a routine bust for Customs and Border Protection's agents. However, upon testing the substance, it was determined to be a highly contagious form of anthrax, and a number of hospitals, decontamination units, and other response elements became necessary.

Interoperable data played an important role in the event, by allowing the production of a common operating picture that compiled data from many information sources. These information sources provided data that were all collected as CAP alerts, or converted from their source data into CAP, and integrated into the common operating picture as feeds. These CAP alerts were delivered to over 150 targeted participants using Swan Island Networks' TIES service. TIES is a situational awareness system providing data integration and display via a web browser. Users receive and view the Common Operating Picture and associated data in a dashboard interface.

Structured information feeds that were aggregated into the COP included:

- **Exercise events.** The Master Scenario Event List (MSEL) had been created ahead of time, and each event was released as an alert as it occurred. Interdiction of the initial ten detainees, decontamination, and 70-plus other events were displayed to participants as the events happened, showing their relative locations on a map and providing information about the event.

- **9-1-1.** Local 9-1-1 incidents were monitored, converted to CAP and displayed on the map, giving participants the ability to see actual incidents going on in San Diego concurrently with the event.

- **Weather.** CAP feeds from NOAA were used to supply any severe weather alerts to all participants in the training event. Rain, hail, tornado, and other warnings were enabled for near-real time receipt and distribution.

- **Earthquakes.** The USGS Earthquake Notification Service provides data about earthquakes as they happen. Had an earthquake occurred, notification would have been automatic and received in near real time through TIES.

- **Tsunamis.** NOAA's Tsunami Warning Centers in Alaska and Hawaii issue tsunami alerts if a major earthquake triggers a killer wave, and would have provided such information had one occurred.

Case Study

- **Traffic status and traffic cameras.** Microsoft Virtual Earth's mapping software provided traffic status for major roads within the San Diego area, and traffic cameras feeds were integrated into the COP dashboard. All could be seen on a single screen, in relation to each other.

- **Remote video feeds.** The event used at least seven Unmanned Aerial Vehicles (UAV) to simulate the real-world flyovers that would occur in a genuine emergency. The UAVs streamed live video feeds to TIES, which represented each feed with a geo-located icon on the COP map. The icons, when clicked, opened a screen to display the live footage.

- **Sensor feeds.** Sensors deployed in the field by ViaLogy and Senusion supplied CAP alerts simulating sensor tests for anthrax and other discharges. These could be added in real time to the COP, giving all participants rapid feedback when a sensor was triggered, indicating a critical event needed attention.

- **Remote location pictures/video.** A key DoD participant executed planned operations in the field, beyond normal event boundaries. Through CAP, the organization was able to transmit information gathered in the field directly into the COP, to be distributed only to certain chosen, trusted recipients. Utilizing a very robust mobile communications environment (provided by CommsFirst), the group also received the full COP, so they knew what all the other participants were doing in near-real time.

- **Alerts from professional providers.** Global alerts from specialized information providers including Global Incident Map and RSOE/Havaria supplied data that was displayed on a world map, providing a global context for event participants. Each alert had a graphic icon in the correct latitude and longitude on the COP map; clicking one would provide detailed information.

- **Hurricane tracking.** A tropical storm in the gulf threatened to become a full-fledged hurricane, and we tracked this storm during the event, getting standardized updates as they occurred. Using these, we created a time-phased map of the storm's path and projected direction that could be shared with participants.

- **Private CAP alerts.** Several groups issued private CAP alerts that were routed selectively to trusted recipients. This illustrates a key aspect of CAP's flexibility: All alerts do not have to go to all people. Security and rules can be applied to ensure that only a selected group of recipients receive a message.

- **Emergency Operations Center data.** Using CAP, the common operating picture itself could send and receive information from WEB-EOC, the software used by San Diego for major incident management. Connecting the city's incident management capability with the exercise Common Operating picture enhanced the overall information picture for all participants.

Lessons learned and validated during the event

- Information sharing can be accomplished across a broad range of participants. The interoperable data approach allowed for significant information sharing during the event. All of this information was shared in near-real time, keeping people up to date and in the loop without extensive conference calls or radio traffic.

- CAP allows "snap together" feeds versus complex integrations. Because of the preformatted nature of CAP, assembling the multiple information feeds into an aggregated picture was easily accomplished, easily tested and validated, inexpensive, and fast

Case Study

to accomplish. Contrasted with the complex, custom data format integrations of a few years ago, this event demonstrated major breakthroughs in interoperable data's capabilities when speed of deployment is important, and the rapid integration of unexpected partners is a critical element of information sharing.

- CAP allows notifications to be transmitted transparently to many devices. In the case of Golden Phoenix, we used a subset (PC, mobile phones with installed software, and SMS text messages), but these alerts could have been routed widely to many more types of devices that understand CAP as an alerting and warning method.

- Structured communications are integral to the management of future major incidents. Hurricane Katrina, 9/11, and other major incidents have demonstrated the need for interoperable and structured communications across the many groups that become involved in a major incident – and the current lack of a cohesive method. This was also illustrated in the 2004 tsunami in Indonesia, where nearly 300,000 people perished from the initial wave and subsequent destruction. Early warning and follow-on communications could have saved many lives, particularly if all the NGOs (non-government organizations such as the Red Cross) would have able to communicate seamlessly, both horizontally and vertically, with each other and with other entities such as the U.S. Department of Defense and the local Indonesian authorities.

Chapter Seven

Interoperable data efforts: things you can watch, use, leverage, and adopt

> *In times of change, learners inherit the Earth, while the learned find themselves beautifully equipped to deal with a world that no longer exists.*
> —Eric Hoffer

As of winter 2010, a number of large-scale interoperability initiatives are in motion. Like all initiatives, some will exceed expectations, several will come up short, and a couple might become outliers in terms of disproportionate progress, getting cancelled or made irrelevant somehow. You also can't rule out stagnation; if an effort loses its champion or sense of urgency, it can end up being stalled for years – even if it is a good idea.

A commitment toward government transparency

During his presidential campaign and as part of his new administration, President Barack Obama advocated that government information should be transparently available to the public and organizations. While this doesn't change the fact that much information will still be classified Secret, Top Secret, and

Silver Bullets

beyond, it does bode well for the use of interoperable data structures as a means of making data from many federal agencies available in a raw format.

Agencies including NOAA have long-standing policies of sharing weather and other information with the public in interoperable data formats. Many other agencies keep information in more closed formats and limit access – often providing graphs and charts or reports, but not releasing raw information. Much analysis has been published on federal web sites, but this limits the consumption to the final product, versus a transfer of raw data. The Freedom of Information Act (FOIA) gave the public a means for trying to pry information out of the federal government agencies, but often it takes so long to process these requests that the useful life of the information is over, or the data would come on paper versus being in a raw, electronic format that could be manipulated.

In 2009, Vivek Kundra became the first-ever Federal Chief Information Officer (CIO) of the U.S.; previously, he had been the CIO of the District of Columbia. In the D.C. CIO role, he opened up many of the district's data sources and made them available to businesses and the public through identified and published data feeds. Maintenance requests, crime reports and a host of other information about the District could all be obtained in a raw XML form by anyone on the web, and analyzed for multiple purposes.

Many data feeds can be integrated into a single web interface. This interface shows incidents of varying types from multiple sources, with a weather radar overlay.

116

Chapter Seven

There are many organizations and people looking to hold the government accountable to its promises of transparency and openness, with the goal of creating a positive feedback loop between the government and its constituents. In early 2009, the Sunlight Foundation sponsored Transparency Camp, an "un-conference" in Washington, DC. This brought together a smart, diverse collection of people, ideas, and opinions; it had a non-conventional format in that the agenda was made up the morning of each conference day, not pre-scripted weeks or months before the event (do a Google search for the "Bar Camp" conference format for more information about participant-driven conferences).

Transparency Camp was attended by over 400 people from the government, non-for-profits, industry, and everyday citizens interested in better information exchange. Many conferences like this one are prohibitively expensive, but this was free. Also, unlike most conferences, it was on a weekend, ensuring that those who attended wanted to be there, rather than have people attending as a pleasant alternative to another day in the office. Topics were very relevant; because of the diversity of the attendees, there was sharp and smart debate. It was one of the most useful conferences I've attended in 25 years of innumerable conferences on many topics.

Another area where the federal government committed to bring openness and full reporting was the $700-plus billion financial stimulus package that was designed to help mitigate the deepest recession since the Great Depression. The goal was to show the full use of the stimulus funds from beginning to end, and to be able to identify the number of jobs created by the government spending. Agencies had to show the initial dollars allocated to them, and all the initial financial outlays to states or other entities.

As these dollars moved down to prime contractors and other businesses, recipients were required to track and show their use of the money including allocations to subcontractors and end recipients. No doubt when the analysis is all said and done, we'll find that some pizza parlor somewhere will have sold a large amount of stimulus-related pies, or we'll be able to compare pizza in Illinois with other comparable food outlays (barbecue, Chinese, and sandwiches) in Texas, Florida, and Massachusetts.

Having the data available in its raw state is important. Various organizations must have the ability to run their own analyses for their own purposes. Some may be concerned only with information from the last 30 days; others may be looking for historical trends. Some of the data may be used now; other analysis might take place 20 years from now, feeding the research of historians comparing current expenditures to a later stimulus package, or to show a return on investment (hopefully positive) on the entire program.

The ability to disseminate these large amounts of information is important to making the government transparent, as are interoperable data formats that can be used by a wide range of organizations. XML and interoperable, standardized formats are the perfect vehicle for this dissemination.

> *Assessment: This is a very strong initiative with excellent people working the problem and committed to success. Like most federal government initiatives, it gets exceedingly complicated as the details are considered, and the people below the executive level sometimes have a different sense of urgency (they may be career employees, not bound by four-year political cycles). In some cases, this is good (minimizing poor, short-term decisions); in others, it impedes progress*

greatly (there is always more time). The transparency "genie" is out of the bottle, and it will be very hard to go back to the days when little information was available in a raw data format. This initiative will increasingly help organizations and citizen watchdog groups to monitor federal allocations and spending, and will be positive in additional ways.

Technology initiatives

The Organization for the Advancement of Structured Information Standards (OASIS) is a not-for-profit organization committed to structured information standards. Founded in 1993, OASIS has more than 5,000 participants representing over 600 organizations and individual members in 100 countries. It is truly a global organization with a broad reach and strong leadership that gets a lot done with a good consensus-based process. OASIS has multiple initiatives under way; three are highlighted below as very important.

1. **ebXML** is a standards-based business process foundation that promotes the automation and predictable exchange of business collaboration definitions using XML.

    "Electronic Business using eXtensible Markup Language," commonly known as ebusiness XML, or ebXML (pronounced ee-bee-ex-em-el) as it is typically referred to, is a family of XML-based standards sponsored by OASIS and the United Nations Centre for Trade Facilitation and Electronic Business (UN/CEFACT). The mission is to provide an open, XML-based infrastructure that enables the global use of electronic business information in an interoperable, secure, and consistent manner by all trading partners.

    The ebXML architecture is a unique set of concepts, part theoretical and part implemented, in the existing ebXML standards work. This work is based on earlier efforts on ooEDI (object oriented EDI), UML / UMM, XML markup technologies and the X12 EDI "Future Vision" work sponsored by ANSI X12 EDI.

The melding of these components began in the original ebXML work and the theoretical discussion continues today. Related efforts include the Object Management Group work and the OASIS BCM (Business-Centric Methodology) standard (2006).

2. **OASIS Blue** is a new initiative aimed at furthering sustainability (typically considered "green") by applying ecommerce methods and focusing on business opportunities, particularly in the area of energy. It leverages several areas in which OASIS has done consensus-based work, in particular transparency, energy and security standards, all of which can help drive Smart Grid and smart building standards. It will look at distributed, interactive energy paradigms that will emerge as more alternative fuels become more mainstream, and many other energy initiatives that will be required for America and the world to maximize the future use of electricity and other energy products.

3. **The OASIS Key Management Interoperability Protocol** (KMIP) technical committee works to define a single, comprehensive protocol for communication between encryption systems and a broad range of new and legacy enterprise applications, including email, databases, and storage devices. By removing redundant, incompatible key management processes, KMIP will provide better data security while at the same time reducing expenditures on multiple products.

Chapter Seven

Assessment: This organization will continue to add value to the world of standards and the world at large. These three initiatives represent only a small part of its worldwide work. If your organization isn't a member of OASIS, you might consider joining or volunteering to help on standards meaningful to you.

The National Information Exchange Model (NIEM) is a government initiative gaining momentum. It began as a partnership of the Department of Justice and the newly formed Department of Homeland Security, leveraging work that Justice had done in creating the Global Justice XML Data Model (GJ-XDM). Since so many terms used in the criminal justice system also exist in other agencies' vocabularies, this has helped synchronize efforts and lay the groundwork for many more standards to come. NIEM has expanded its domains to include Justice, Intelligence, Immigration, Emergency Management, International Trade, Infrastructure Protection, and Information Assurance, with others likely to join the initiative.

Assessment: NIEM is grossly underfunded for the impact and value it could have on streamlining government, making data more transparent and usable, and coordinating usage among federal, state, local, tribal, international, and private organizations. It has made continued, incremental progress over its existence and will continue to work this very complicated problem – but changing the government is hard and expensive. The key to success is scaling the number of transactions that flow using the NIEM formats, and evangelizing how important interoperable data is for future information sharing efforts across all organizations.

Weather Info for All (WIFA). Kofi Annan, former Secretary of the United Nations, announced this initiative in June of 2009 at a conference in Geneva, under the auspices of the Global Humanitarian

Forum. Climate change is responsible for some 300,000 deaths each year and over USD $100 billion worth of economic losses; sub-Saharan Africa accounts for close to a quarter of these losses. WIFA is being implanted to radically improve Africa's weather monitoring network in the face of the growing impact of climate change. The World Meteorological Organization (WMO), Ericsson, and a host of other partners will also contribute to the project. While just under way, this initiative's goals are far-reaching, yet seem very achievable. Progress means the saving of lives and property, and the creation of an information channel that could be used for many other worldwide humanitarian causes.

> ***Assessment:*** *This is a major initiative, and will make rapid progress if the founders get the right sponsors and support. The end result will be extremely valuable and noteworthy. WMO has been at the forefront of the use of Common Alerting Protocol, which should allow weather alerts in real time to reach the furthest corners of the globe in a format that will be timely, usable, scalable, affordable, and archivable for future research.*

FEMA and CAP. The Federal Emergency Management Agency (FEMA) has been working to adopt CAP for the last several years. CAP can be used in the Emergency Alerting System (EAS), the Integrated Public Alerting and Warning System (IPAWS), and other initiatives – both proactively and after a disaster occurs. CAP version 1.2 has several FEMA-specific modifications that have been recently adopted.

> ***Assessment:*** *I hope FEMA can get their initiatives moving now that CAP 1.2 has been approved. It's unclear why things move slowly there, but the agency has been beaten up by many of the*

Chapter Seven

blue ribbon panels appointed to figure out what went wrong in disaster response. Information standards such as those that NIEM is developing will be one major improvement for FEMA to adopt and implement. Your life might depend upon the outcome, so this is one you should pay attention to.

Sensorpedia is a program initiated by Oak Ridge National Laboratory (ORNL) to utilize Web 2.0 social networking principles to organize and provide access to online sensor network data and related data sets. Sensorpedia is based on the same underlying social networking and collaboration principles used by popular web sites such as Wikipedia, Squidoo, Google Maps, and Facebook. Instead of networking users based on mutual personal interests, Sensorpedia networks users based on mutual sensor information interests. It provides near real-time collaboration among communities with requirements to share sensor information. An open API and flexible access controls ensure Sensorpedia will work for everyone, regardless of application requirements.

***Assessment:** This is a great concept. Open repositories like Sensorpedia will help increase the amount of data sharing and interoperable formats, and will stimulate the number of tools that are available to work with these data sets. There will be many of these different, specialized sharing areas, and my prediction is that they will slowly consolidate, or a very strong search capability will develop across the various sites, allowing users to pick and choose from published data sets.*

Smart Grid interoperability standards, from the National Institute for Standards and Technology (NIST). Under the Energy Independence and Security Act (EISA) of 2007, NIST is tasked with organizing and managing the large number of standards that must be implemented to realize the vision of the Smart Grid. The vision entails revamping our energy grid to accommodate new and renewable energy sources, increase security, and make energy use more effective and efficient. Communications between all the players and many types of devices will need to be greatly enhanced and made compliant with today's and tomorrow's technology platforms.

> *Assessment:* *The urgency of this initiative will help drive standards into production quickly. Multiple billions of dollars were released into Smart Grid projects in 2009, and these projects will help drive the need, field-test resulting standards, and allow rapid iteration and improvement over the coming years. There are many players, and the task is formidable. My prediction is that the vast amount of money being spent will help drive the standards process forward as both utilities and vendors realize that (as with the Internet) nobody is going to adopt a proprietary technology grid-wide.*

The Unified Incident Command and Decision Support (UICDS) is a middleware framework being developed by the Department of Homeland Security (DHS). SAIC, a large systems integrator, is working to construct this innovative capability, and is coordinating the interface with technology providers and government agencies. UICDS will allow information to move from one application to another using standard digital formats. It will not supplant the creators or organizers of content, nor operate as an application for visualizing information. Rather, it will function as an organizer and coordinator between the very large number of applications that have to be connected in order for DHS and the country to get an accurate, blended, comprehensive picture of what has happened, is currently transpiring, and might (through predictive analytics) occur.

Chapter Seven

> *Assessment: This initiative has a very good chance of succeeding and helping to "plumb" the government and all its respective partners for emergency information sharing, because it uses the National Information Exchange Model to the maximum, incorporates other federal and commercial standardization efforts, and works closely with commercial developers.*

eXtensible Business Reporting Language (XBRL) will enable financial transparency by Fortune 500 companies. The Securities and Exchange Commission (SEC) will be mandating the use of this global standard for financial reporting. Early voluntary programs have shown vast cost savings and increasingly improved accuracy – as well as the standardization benefits when comparing one company to others. Edgar Online has excellent visual descriptions of what XBRL is and why it is important.

> *Assessment: Cultural and non-technical factors come into play when adopting standards, and these issues will factor heavily into adoption in this industry. Some companies don't like transparency – obfuscation is much preferred. This standard makes so much sense that I predict it will overcome all these obstacles – though I think it will take regulation and mandates, versus just common sense, to have it adopted nationwide.*

Health Level Seven (HL7) was begun in 1987, long before XML showed up on the worldwide scene. It is an accredited standards body whose mission is to rationalize and standardize the myriad of heath care information entities that exist today. It is an initiative that bears intense scrutiny.

> *Assessment: My opinion is that if anything is going to cut healthcare costs and improve care over time, it will be accomplished through standardization efforts that will in*

125

turn reduce overhead and waste, and allow for meaningful comparisons. Similar to our shipping container example earlier, the right standards could substantially improve the information plumbing of the health care system. This could give the visibility necessary to help determine the right policies for allocating our health care dollars as a country.

Social media. While it's not a specific initiative from a specific group, social media applications are sweeping the world, and becoming some of the most widely used platforms for communications in human history.

Social networking is a collection of ways of staying in touch with all your friends. In social networking, "friends" are loosely defined, and can evolve very quickly to followers and other individuals that have a major or minor interest in you or someone connected to you. Two of the major consumer social networking capabilities are Twitter (www.twitter.com) and Facebook (www.facebook.com), both of which have many millions of users. In the business community, LinkedIn (www.linkedin.com) is the current predominant networking tool on the web.

Each of these tools has evolved since its inception, and all have added structured data and interoperable elements as they have grown in scope. So that you get an understanding of how structure can help, I'll touch on a few standard data interoperability items that have evolved in each.

- **Twitter** would at first appear to be the most unstructured of applications. Each Twitter user can "tweet" or broadcast very short (140 characters maximum) messages into the Tweetosphere. Anyone can follow anyone (although it's possible to block undesirable followers) to tap into the

Chapter Seven

Tweeter's stream of consciousness, which could range from what they are wearing and where they are going right now, to existential thoughts as they wander down a city street or mountain pathway. The amount of Twitter traffic has become enormous: U.S. traffic to Twitter grew 1,382 percent between February 2008 and February 2009 - from 475,000 unique visitors to 7 million. One key feature that's evolved in Twitter is the "hashtag," a structured identifier used to segregate and identify tweets around a specific event or topic. For example, "#crisiscamp" might be added to every post that I do while at a particular conference. This small amount of structure allows everyone in the world interested in these messages to follow that topic. Another example of Twitter using structure is bit.ly, a compressed address scheme for another website page address, which is abbreviated to fit within the character limitations of a tweet.

- **Facebook** allows friends to stay in touch with each other by pasting, writing, commenting, and a host of other means of interaction on the web. Everyone gets a "wall" which is interlocked through relationships and permissions to other people. The site uses a host of interoperable elements to enable blending text, photos, videos and more into a wall. In addition, there is structure that allows you to search for people, and the structure is smart enough to suggest people that you know as well.

- **LinkedIn** is a very good business networking tool. I've been a LinkedIn account holder for several years, and it's a comprehensive way of connecting with people I knew or interacted with in the past. I can also request introductions to people I might know through a few degrees of separation. The site has both free and paid users, depending on the capabilities and access that people choose. Like Facebook, LinkedIn's interoperable, structured data elements enable people to give others permission to connect based on whether they have done business together, worked at the same company, attended the same school, or are friends. As you build your profile, you enter information that can be linked to other people automatically, suggesting people that you may know. The site uses interoperable data to interact with other partner sites such as Slideshare.net , which is a repository of Microsoft PowerPoint presentations (and other formats) on almost any topic. The two sites can share information, and presentations I upload to slideshare.net are suddenly visible to my connections on LinkedIn.com. If you are a business person, get yourself an account (free) and send me a connection request, and you'll see how easy it is.

Assessment: *As more data becomes structured, you'll see more transactional and structured functions in social networking sites. For example, you might be able to have your iPhone update your geospatial location on a schedule. That structured transaction could trigger analytical processes that result in others knowing your whereabouts. Suddenly your friend realizes you're nearby and asks if you want to have breakfast. Or you get an alert about a radar trap from other nearby user. Or, if you allow it, a marketing offer (for something you've indicated you're interested in) from a nearby store.*

Chapter Seven summary

There are a number of important, world-changing initiatives going on that are dependent on a good set of interoperable standards. If you pay attention to the developing capabilities, you'll be better able to determine what's useful to you or your organization.

1. Many factors will affect the impact, speed, and ultimate success of these initiatives – the technical plumbing is easier than the sometimes gnarly policy issues surrounding many of them. Despite obstacles, expect many of these initiatives to take hold.

2. Increasing progress has been made by organizations adopting XML-based standards, which bodes well for future adoption and expansion by others. We're at the beginning of the growth curve.

3. The information standards road is becoming a highway versus a byway, and will move toward becoming the accepted method for exchanging important, cross-organizational information.

4. The impact of these standards is global, which is important to consider in design, deployment, and utilization. XML is being used around the world, and several standards developed for U.S. usage have become globally accepted.

5. The private sector will continue to become more important in driving the adoption of external standards. As of this writing, large Fortune 500 companies have almost no comprehensive and standardized external information on incidents, alerts and other elements that can impact their employees and operations – because every jurisdiction they deal with is different. Demand will drive standardization.

Chapter Eight

Executing a data interoperability pilot (*right now*)

> *We are what we repeatedly do.*
> —Aristotle

A bias toward action

It's time to get out of your chair, off the couch or the treadmill, and start moving forward into the future. Let's make something happen!

The key to any long-term journey is taking the first step, coupled with the willingness and courage to disembark from where you are now (which may be perfectly comfortable). Managing risk is important; anything really new will have a higher percentage of failure than the tried and true approaches of the past – unless you wait too long, until the old ground is crumbling underneath you. If your old world is falling apart, moving rapidly to a new paradigm may be your only choice, and one that you had better run toward, even if you are carrying scissors.

So, who to send on this future-oriented journey? Who do you trust with your organization's future? An old hand who's seen it all? Or a new player who has everything to gain and nothing to lose by advocating change? How about one of each?

Each organization develops its own cultural methods for injecting innovation, change, and trying new things – top down, bottom up, executive champions, and "skunk works" are only a few of the approaches that have emerged over the history of man and technology. Innovation is very difficult and best left to be nurtured by a small group, versus a large, multi-departmental team with a host of conflicting agendas. Attempts at innovation with mature, global analysis can take place concurrently with the wild, hair-on-fire approach, but each has a different purpose that should be identified and followed, and each needs its own dedicated resources that have separate stakes in success.

Once an innovation is proven or deemed worth the risk, deployment in large organizations is still fraught with difficulty, conflicting priorities, and budgetary issues. You may need it for your organization in six months, but by the time you develop, deploy, train, retrain, re-launch, and succeed – it might have taken years.

Plan your strategy for 100 years, but start the pilot right away

There's a famous story of a French nobleman who had a vision for a row of mighty oak trees lining the carriage entry of his country estate. He summoned the gardeners, and they started protesting that it would take 50 years for the oaks to grow. The nobleman was not swayed – his take on the situation was that it was even more important to get them planted that very afternoon if the trees were going to take so long to grow! The same should hold true for your move toward interoperable data.

If you are part of a large organization, you will likely have an enterprise architect and/or a data architect tasked with management

Chapter Eight

and strategic planning of the organization's historical, current, and long-term data. This person or group is often isolated in the information technology department, and doesn't get the full flavor of the organization's needs for the future, or the importance of making current data collections more available, accessible, and usable for unexpected and highly strategic purposes.

Many organizations will give functional groups such as sales or marketing extracted data summaries that are quickly outdated and out of sync with recent data, yielding conflicting answers and in some cases, creating more havoc than help for the situation under study. In other cases, system transitions can remove much of the intelligence of historical data, and cause massive reformatting and re-engineering issues. It is all too easy to throw away or neutralize vast quantities of data that could be valuable in the future.

Having control of your information is an important element for every organization, and it's vital to understand which information will be strategic to your organization's future, versus something that will be available just for operational purposes. An increasingly important factor in your strategy will be interaction with other, external organizations. It may be useful in the future to share data that you may not be able to, or see reason to, share today.

As an example, a coalition of retailers might want to share information about checking account scams or identity theft between their security groups, without revealing how well their respective specials of the week are doing, or the identities of their customers. Or a major manufacturer might share detailed field maintenance reports across a wide range of suppliers under strict confidentiality guidelines.

Taking stock for a pilot project

There are many different options that you can utilize for an interoperable data pilot. Here are a few things to consider:

- **Scour the web** for ideas and information. Your needs and innovation might not be so unique that someone hasn't already figured it out. If you can learn from a pioneer who already knows the way, do it. It is a far better approach than making known mistakes all over again. You will find that most organizations love to share innovations they've made (you wouldn't have found them if they were trying to keep it a secret). There are a lot of potential ways to locate new found knowledge. My first choice is to start with Google, looking for PowerPoint presentations and Adobe Acrobat documents using the advanced search options.

- **High payoff on success.** If you have a couple of potential projects that will deliver real benefits to your organization, that's a great place to start; everyone will pay attention, there will be a sense of urgency, and resources can be consumed without a fight. A major success that gets the proper recognition can pull an entire organization forward and lay the groundwork for a new generation of capabilities.

- **Minimal risk.** You don't want to bet the farm and the organization's mission on something new, so look for ways to keep the exposure to a minimum in the event of setbacks or some kind of outright failure. This seems like common sense, but many organizations have put untried technologies and approaches on their critical path unnecessarily, and paid the price for failure.

Chapter Eight

- **Fail fast.** Find a starting point that can prove out the capabilities and vision you determine in a short period of time – days if possible, weeks at the most. If the project doesn't take hold, you can see what went wrong and reset the course on the next attempt. Keep repeating this process, and revitalize your efforts. Reward the people willing to risk failure – they are the ones who will be your future leaders. Don't believe that you have to spend months or years before getting started – it doesn't work in today's world.

- **Outsource** through an external service. The capability you are considering might be available immediately as a managed service (perhaps on the web) by a company already servicing multiple customers with similar needs. There are a host of issues requiring due diligence (security, stability, scalability, redundancy, and more), but this approach might jump-start your efforts significantly. This approach can also greatly reduce the cost of a pilot project; services can typically be cancelled quickly, require no added hardware, no software (if the service is web-based), and can be set up very quickly and effectively. Under these circumstances, the development risks have already been taken and resolved by someone else. No company can better illustrate this than Salesforce.com. This company can provide a turnkey solution to Customer Relationship Management in an afternoon, versus the 18-month internal evaluation and implementation process it will take if you do it yourself.

- **Collaborate** with non-competitive partners. You may have an industry association that allows you to interact with companies that have missions similar to your own, but are non-competitive. Since interoperable data is really about sharing, finding an external entity to share costs and risks and help evaluate solutions may be very viable. Even competitors can share non-competitive data – severe weather, notifications of new fraud scams, and other interests serving the common good of the industry or community.

- **Evaluate other attempts.** Find a way to interact with companies starting comparable projects and look for best practices, problems that may already have been encountered, and a successful route. Rarely will you be the only one traveling this new road – the world has become so multidimensional and interconnected that finding others trying to solve the same problem will be much easier. Try to find the similarities and differences, and use them as constructive feedback to your own efforts. Be willing to share with others as well.

- **Avoid a purely technical project.** The project should be driven by a champion who understands the business value, the technical underpinnings, and the ultimate vision of what could be accomplished. This may require an executive sponsor who can keep the project visible for senior management and the board of directors, but having a high-level champion will avoid the silo mentality that can happen when a project is focused too narrowly.

- **Acquire standardized data feeds.** One low-risk way to experiment with interoperable data is to bring in outside data feeds and utilize them as an interoperable format. An example of this would be to create a 360-degree threat picture composed of feeds about weather, traffic, 9-1-1 incidents, and power outages around geographic areas important to your organization. This information might all be acquired in

Common Alerting Protocol format, and you could perhaps augment it with your own internal data.

Some things to avoid

I've spent years managing various projects and innovation efforts across a wide range of companies, and you can learn from the mistakes I've made and observed. This isn't an all-inclusive list, and there's plenty of room for new, innovative ways to fail – though I have found that a new way of failure, despite the pain, is still far better than repeating the mistakes of the past.

- **Using the wrong leader.** When there is a new project, and someone has to be appointed, many organizations pick the person with nothing to do – usually a bad idea. The leader you really want is the person who, no matter how many important initiatives he has, can always find a way to do more and accomplish more.

- **Betting the farm.** Putting your entire future on one project is not a good idea, but every year many organizations get optimistic (or desperate) and end up doing just that. There are always one or two horror stories about runaway systems projects that cripple or kill companies, even those that have been around for a long time and should have known better. New approaches are of little use if you are out of business.

- **Hiring consultants to run the project.** If the project is worth doing and the idea is to establish a baseline for the future, you're better off building the expertise inside your own organization. Hiring consulting expertise that can "teach you to fish" is highly recommended, but avoid renting a fleet of boats to do the fishing for you! If you want to inject fresh thinking, hire a few college or graduate student interns from your local university – you'll be surprised how much energy and new thought will result if you pick the right candidates. I hired Justin Buckley, one of the reviewers of this book, right out of school, and he was an invaluable team member in no time. You'll also create a pipeline of people to hire if the project is successful.

- **Inflating ROI (return on investment).** For early projects, many organizations require a complex return on investment calculation—which turns out to be based on unknowable, imaginary numbers, or inflated. Imagine Queen Isabella asking Columbus for a detailed investment analysis before he executed his general, directional strategy of sailing west to go east. He might have provided it, but he likely wouldn't have believed it, and wouldn't have accomplished it either. ROI often comes in surprising, unforeseen places.

- **Failure to create success metrics.** You'll never know when you're done if you don't have clear metrics for success or failure of the project. This should be more than monetary expenditures and completion date – what you were trying to accomplish should be clearly understood before you go forward. You might end up making mid-course adjustments and revamping the goals, but setting specific, measurable objectives is a vital element for success.

My pilot succeeded, now what?

Replicating the results of a successful pilot throughout the organization can be a pleasant process or an utter disaster, depending on the approach, the timeline, the budget, the people, and a variety of other variables.

Chapter Eight

It is important to remember that innovation is difficult, and that almost all innovations will change someone's world inside the organization. Making sure that the proper groundwork is laid for the change is an important element of any major shift in strategy, technology, or manner of doing business. Moving a new idea into a massively larger environment may uncover vulnerabilities or problems that weren't discovered in the pilot, and contingency plans should be in place to allow for rework and adjustment.

It's key to identify the champion for the change being implemented. If a business change is being driven by the information technology department, the odds of failure go up, regardless of the good intentions of the implementing group. There are many useful books on technology project management and the social elements of change that can help. You must remember that ignoring the people issues is a classic recipe for failure.

Silver Bullets

Chapter Eight summary

1. Get in the pool! Nothing takes the place of getting in the water and trying to swim.

2. Innovation is hard: Executing a new way of thinking within most organizations is difficult, and it seems to slow down in direct proportion to the size of the company. Do it anyway.

3. There are many ways to approach innovation. It is important to look at what has worked previously, and apply those findings to your organization. Having concurrent and competitive efforts may be an approach to try.

4. Early wins: If you can establish some early gains and build upon them, this will help you move forward into larger and more profitable projects.

5. Cross-organization involvement: These type of projects demand collaboration between various groups in the organization, but can be a viable proving ground for information exchange. Look for other data sharing zealots to help champion your efforts.

6. Learn, debrief, share. These types of projects generate a lot of new learning and will stretch your organization. Be sure to take time out and assess what was useful and what wasn't. Replicate the things that work.

7. Keep trying: You must have a strategy, so even if you encounter a few early failures or shortfalls along the way, your sense of direction will allow you to get back into the game and keep trying. Learn from your mistakes, keep trying, and the journey will be worthwhile.

Case Study

External alarm interface exchange standard

Here's a case study in which a standard has been completed, tested, and issued to the world. It's complementary to Common Alerting Protocol; it's an XML standard; it's been through strong consensus building; and it's poised for implementation. This standard can save lives and property, reduce government costs, and aid first responders. It underscores the value of an interoperable data standard for a specific task, and illustrates the many benefits received when it is deployed. Thanks to Bill Hobgood of the City of Richmond, Virginia for all the material regarding this effort.

9-1-1 Public Safety Answering Points (PSAPs) have a demanding, 24x7, real-time set of demands. They are the first to be called when bad things happen, and they provide life-saving dispatch services for police, fire, and medical services. The model for this until now has been mostly manual; people dial 9-1-1 to reach the PSAP to report an incident or request help.

Automated alarms for security, fire, and other emergencies place heavy demands on a PSAP. A typical scenario: An alarm placed in a home or business is triggered by some kind of event. It notifies the alarm monitoring company to which it belongs. A human operator working for the alarm monitoring company then manually calls the PSAP to give initial details, and continues to communicate with the PSAP if there is a legitimate incident. These calls add significantly to the call volume of the 9-1-1 center, and add two or three minutes to the time it takes the PSAP to process the incoming information. Two or three minutes can be a long time when a life is at stake.

Silver Bullets

In 2004, the Association of Public Safety Communications Officials (APCO) and the Central Station Alarm Association (CSAA) jointly identified a need for an automated solution that would notify the PSAP, allow bidirectional communication between an alarm monitoring company and the PSAP, and enable directional updates of other events related to the open alarm event. In January of 2005, APCO and CSAA announced a formal partnership. York County, Virginia, was selected as the first beta site, and Vector Security as the CSAA member company participant. In July 2006, the capability went live. In August of 2006, the City of Richmond Police Division of Emergency Communications project went live as well. Fire and medical alarms have been added, with very favorable results.

After three years of operation, the capability has handled more than 7,000 alarm exchanges between Vector Security and the two Virginia PSAPs. Benefits include:

- Over 7,000 fewer calls to the PSAPS
- Elimination of verbal miscommunications between the PSAP and alarm monitoring company
- Significant decrease in response time – yielding more apprehensions (police), fires extinguished (fire), and lives saved (medical)

The standard has become an ANSI standard and has been adapted into a NIEM IEPD (Information Exchange Package Description). Thanks to all who help pioneer this life-saving standard.

Commentary

The implications for savings and improved service are huge. If all calls were moved to this method of answering over coming years, it's estimated that up to 32 million calls from alarm monitoring companies to PSAPs per year could be eliminated, in the U.S. alone. With alarm monitoring system prices going down, and the number of installations going up, the number of events that PSAPs must handle will only increase over the coming years.

Case Study

This standard can also lay the groundwork for other automated standards to be implemented for PSAPs. These centers are always there for everyone when emergencies happen, ready to coordinate life- and property-saving response. The people who run them are committed, capable, and hard working public servants who don't get near the recognition and thanks deserved. I attended an awards banquet for the 9-1-1 team in Portland, Oregon, several years ago, and still remember choking up when the awards were handed out – one for someone saving a baby's life by teaching the caller CPR over the phone, and another helping to deliver a baby over the phone (The Stork Award, as I remember).

Chapter Nine

The far-flung future: where do we go from here?

> We can't solve problems by using the same kind of thinking we used when we created them.
> —Albert Einstein

> The Web as I envisaged it, we have not seen it yet. The future is still so much bigger than the past.
> —Tim Berners-Lee

We're moving at light speed in so many directions – people, technology and more. In this chapter, we'll consider several of these issues and look at where interoperable data and its descendants, combined with other technology trends, will be important.

Global trends

The state of the world has a huge impact on the development and implementation of interoperable data standards. Here are a few issues to consider. It's well to remember that while each has its own considerations and implications, it is the interaction between them that will generate unpredictable consequences – for both the good and the not-so-good.

1. **More people!** The world's population is continuing to increase, especially in developing countries around the world. This may slow down or flatten over the coming years, but for now, it continues to put more pressure on the planet just from a sheer number-of-bodies standpoint. And more people are at risk. Even simple interoperable programs like severe weather alerting can save lives and minimize damage.

2. **Developing world.** The developing world wants what the developed world has – lifestyle, education, transportation, travel, food options, communications, safety, and security. On the physical demand side, this pressure is higher than mere population increases; everyone's footprint becomes bigger and consumes more natural resources, energy, food, and other commodities. Interoperable data will continue to improve supply chains and distribution around the world.

3. **More Internet users!** The Internet has grown spectacularly over the last 25 years, but still reaches a minority of the world's people. Technological advances on every front and ever-decreasing communications costs are enabling connectivity growth everywhere. At some point in the next 100 years, having access to the world's information services will be as pervasive and necessary as having clean water and adequate food (both still big challenges for large parts of the planet).

4. **Mega-cities.** The world is centralizing, and in 2008 crossed a major threshold: More people now live in urban areas than in rural areas. Cities of over 10 million people are becoming more common, and are projected to increase dramatically over the coming 50 years. This population density exacerbates the issues of how to deliver good quality of life on a normal

day – balancing transportation, crime, jobs, food, energy, and all the other variables necessary for a quality existence. In addition, a 20-million-person city in crisis is a very large problem – and many of the population centers such as Mexico City, Tokyo, Mumbai and Sao Paulo are located in earthquake or severe weather zones, prone to tsunami, tornados, and hurricanes. Interoperable data can help the various silos of information within these cities be shared, helping to optimize daily life, and aiding survival in the event of an emergency.

5. **Technology reaching the poor.** As technology matures and becomes more of a commodity, we're starting to see it reach the lowest levels of society. One cell phone might be shared by 15 people in an African village, but the trend toward connectivity is accelerating nonetheless. The cost of technology has been falling relentlessly, and there is no reason to believe connectivity will decelerate over time, especially as adding more capacity to text messaging, video and other social networking methods continues to advance. Interoperable data will help standardize the flow of future information.

6. **Global pandemics**. It's been estimated that a contagious disease such as the Spanish flu of 1918 could travel around today's world in 18 hours, versus the 18 months it took to spread back then. This has huge economic, medical, political, and emergency management impacts that can be simulated, but are not understood nearly well enough. Real-time communication and collaboration on emerging health threats can be greatly enabled by systems using interoperable data.

7. **Global warming/climate change.** We're all aware of the problem, but many of us have seen minor inconveniences

versus major disruptions, as of winter of 2010. I'm of the opinion that this problem is advancing more rapidly and is cumulatively bigger than many scientists have predicted, and that we'll start to see large consequences. Collection and dissemination of data around this threat will be simplified as standards emerge for collection, analysis, and historical archiving of events.

8. **More complex disruptions and risk.** As interactions and complexity between organizations and countries climb and cities increase in size, there will be new and more complex disruptions that will affect people locally and internationally – and at an increased speed. If a city that's the sole supplier of a worldwide commodity is suddenly quarantined, the effects will quickly be felt throughout the world's supply chain.

9. **Information speed.** Bill Gates, founder of Microsoft, wrote a book a few years ago titled *Business @ the Speed of Thought,* in which he predicted that a burgeoning number of the world's citizens would have IAYF ("information at your fingertips"). His predictions have become reality and will accelerate – more information, more people, faster distribution, and filtering of the information to get the right information to the right person without overloading people from a mental or device standpoint (reading a 500-page document on a cell phone is still not a very attractive prospect).

10. **Energy crises and disruptions.** We have seen spiking gasoline and other energy prices ripple through world markets, changing the habits of car buyers and energy-conscious consumers. Blackouts in South America caused chaos in 2009. Most projections are that these crises will continue to happen as the developing world tries to climb to developed-world status.

Chapter Nine

This could induce major change into the world's growth and expansion equation. Numerous initiatives such as the Smart Grid bode well for better utilizing the energy we have, and many alternative energy sources are being identified and developed. One problem is that growing demand is outstripping technical progress; this may have devastating consequences over the coming years. Interoperable data can enable the exchange of information between the electrical producers, distributors, and consumers – even to smart machines, such as a refrigerator that regulates its own energy use.

11. **Non-state terrorism and crime.** Disruptive organizations will continue to foster major events of destruction against the developed world. Organized crime that spans countries becomes more commonplace every year, creating a major threat to countries such as Columbia and Mexico. Information sharing between local and worldwide law enforcement is a critical need now, and will continue to increase in importance.

12. **Rogue states.** We've seen North Korea and Iran both rattle the sword of nuclear weapons, and it will likely be a problem that continues to unsettle the world.

13. **Interconnectedness.** Whether you like it or not, you are very dependent on many other organizations and countries for your quality of life. A simple example is a power outage: Life changes dramatically when you're back to flashlights, candles, and cold water. Chips are made in China, bananas come from South America, cities have only a three- to five-day food supply, and the government tells you to prepare to be on your own for 72 hours in an emergency. This level of interdependence is going to continue to increase.

14. **Things we can't imagine.** If this were 1910, would many of us be able to predict what was coming in the next 50 years? One breakthrough by an individual or a group of individuals can have almost instant impact on the world at large. Let's hope for some great new things that benefit humanity greatly.

Technology trends

There are many technology trends percolating throughout the world that will have a significant impact on our lives in the next year, five years, ten years, and beyond – all with a relationship to interoperable data. Some are coming to maturity and starting to reach mass adoption, others are just getting started, and still in danger of being overly hyped – will we ever get true voice recognition? Or will artificial intelligence make a stunning breakthrough after all the years of derision?

1. **Competition.** There is a renewed competitiveness in the technology world, currently driven mostly by Google, but augmented by thousands of other companies. Google is developing phones, operating systems, GIS capabilities, and search at a very fast rate, challenging the status quo on many fronts. The worldwide recession has caused small companies to find innovative ways around traditional financial barriers, and I predict that several very strong, world-changing companies will emerge out of the current chaos.

2. **Sensors.** Thirty years ago, sensors cost thousands of dollars and had to be hard-wired into local computers, or their data had to be collected manually. Today, we're accelerating toward sensors that cost pennies each and can be wirelessly accessed from many different points. Imagine never losing your keys,

really knowing if you need milk at home when you are at the store, and thousands of other uses that will develop from these devices moving into the mainstream. Almost all these sensors will communicate via interoperable data transactions. The milk sensor has been a standard joke throughout the technological industry; it will be interesting to see if we ever get there.

3. **GPS and geo-location capability.** The first Global Positioning System cost the equivalent of billions of today's dollars – satellites needed to be launched, complex electronics installed, and a highly classified set of protocols enabled. Today, most phones contain a GPS detector, and mobile GPS with tremendous capabilities can be purchased for under $200. Photos can be geo-tagged by a camera so that you know where they were taken, and people and equipment can be tracked in real time across a wide area. Personal locator beacons and devices like findmespot can be activated when a crisis happens, allowing assistance to be provided to an exact location versus a wide-area search of a mountainside or forest.

4. **Internet Protocol Version 6** (IPV6). Every device on the Internet must have an Internet Protocol address in order to function. IPV4, which has been the basis for the huge growth of the Internet, has to be "tricked" in several different ways to allow this, because the number of addresses is too small for today's burgeoning number of devices. IPV6, when fully adopted, will allow for a much larger pool of addresses, additional security capabilities, and other advances that will make the Internet more usable, more scalable, and more secure. Because of the massive installed base, this conversion is difficult and time-consuming. Much like the electrical grid,

the success of today's Internet has made it harder to upgrade entrenched infrastructure, and it's become increasingly vulnerable to attack as it has grown and expanded.

5. **Mobile phones.** The number of mobile phones is growing far faster than the number of personal and business computers. New, advanced phones combine computing and communications capability seamlessly. In developing countries, mobile phones have preempted the development and deployment of the land lines that support traditional phones, allowing those countries to leapfrog ahead of the countries that have extensive wired infrastructure. The iPhone from Apple and the Android phone from Google both illustrate the direction these devices are going, and the tremendous new capabilities that are being developed around the world. So many successful applications are being developed by third parties, often small startups, that a market for stand-alone mobile app development was created. As one example, PointAbout provides sophisticated development services for mobile phone applications that take advantage of most new capabilities being installed on each new generation of smart phones. Many of the applications built for smart phones will communicate via standardized data transactions, and currently available information feeds can be accessed almost instantaneously.

6. **The Smart Grid.** The global electrical system powers everything, and is a key element in the functioning of the world. However, the grid is under cyber attack, groaning under its own growing weight of use, and tasked with encouraging and incorporating alternative energy. To produce more energy

Chapter Nine

more efficiently, to support urbanization, population growth and economic acceleration while maximizing conservation and storage, the world must move to a Smart Grid – a smart power grid that communicates with all its variously-owned parts and other infrastructure such as smart buildings that regulate their own use. An effective Smart Grid will require much information sharing, all enabled by automated interoperable data transactions.

7. **Cloud computing** (managed services and utility computing). Computing has started to come full circle. In the early days, terminals were attached to mostly IBM mainframes, which processed all the data in highly controlled environments comprised of raised floors, air conditioning, battery backup systems, and extensive fire protection systems. With the advent of the personal computer came local applications, local hard drives, and freedom from the highly structured environment of the data processing group. Today, many applications are moving back to a utility, cloud, or managed service model where the information is all stored on a private network server on the web (and ultimately in an environment that is protected, sterile, backed up, with redundant power). Salesforce.com, for example, has eliminated the installation of software at either the personal computer or even the corporate network level; all software is accessed over the web through the Salesforce servers. This model allows for more rapid development, faster and cheaper adoption, greater security, enhanced information sharing, and many other advantages. It also makes security more important, and connectivity even more essential to everyone's ongoing activities.

8. **Open source.** Open source software is starting to grow rapidly at the applications level, and companies are beginning to form around the idea of a free software suite supported by training, professional services, integration services, and technical support services. This is a very powerful trend that could upset the balance of power for traditional application behemoths like SAP and Oracle. Increasingly, open source software which can handle many basic functions is being released, and there is a growing population of community-minded developers who are working on making such projects more valuable. Google announced its Chrome open source operating system in November 2009, which created an enormous amount of excitement in the world technical community.

Big hairy problems

Several really big, overarching problems need to be solved in order for the world to move into next generation of applications and capabilities. Interoperable data transactions will help in some of these areas, but the combination of growth, complexity, flexibility, and the ever-increasingly important role of computing will keep the pressure on, and make these problems very difficult to solve.

1. **Identity/authentication.** One of the early cartoons about the Internet was, "On the Internet, nobody knows you're a dog." On a computer that was confined to a physical space and owned by one organization, identity could be tightly controlled. But once you connected millions or billions of users with millions of back-end web sites and servers, identity became a very sticky problem. The traditional username-and-password security is still used in most instances. People can end up having 50 or so separate account combinations of different passwords –

Chapter Nine

which need to be changed at different intervals and at different levels of "strength" (complexity of numbers, digits, password length, and special characters). Unsurprisingly, this results in people writing down all their passwords and keeping them next to their computer. Obviously, this defeats the entire purpose of having secure passwords. Other schemes are being tried – biometrics, multi-factor authentication schemes using key fobs and other unique devices, etc., but no method has been developed that can handle the comprehensive needs in today's computing environment. Ultimately, this information should be federated within an interoperable framework, with appropriate protections and safeguards. There may have to be different formats for casual security (a fly fishing discussion site), medium security (your email and LinkedIn passwords), and high security (your bank account). Keep your eye on this one – your iris literally may be one of the identifiers you will need.

2. **Privacy.** When you interact with hundreds of different web sites in the course of a month, you generate a tremendous amount of information about yourself. Used without your permission, this information can subject you to massive amounts of marketing analysis, unwelcome solicitations, and a very Big Brother set of data that you might not wish to share with the world. Some progress has been made in this area, especially by organizations such as TRUSTe, an association of privacy-oriented individuals, organizations, and non-profits. TRUSTe has cajoled organizations into declaring their privacy policies to the public via their sites, and enabled certain watchdog capabilities. The problem gets more complex when you start dealing with jurisdictional boundaries of countries – Europe

has more stringent rules than the U.S., some countries have few rules at all. Certain types of activities can cause multiple problems; adult material and gambling are two examples. Protecting the privacy of minors is very difficult, particularly if a child is using his parent's credit card and behaving like an adult. Interoperable formats can be useful, by allowing a customer to declare how she chooses to be treated, and how she is willing to have her information shared.

3. **Cybersecurity.** If you've had your identity stolen or had spyware take over your computer, you're very aware that the overall security of the Internet and most computing environments is a continually evolving, increasingly complex struggle between the good guys and the bad guys. Individuals, organizations, and countries can be responsible for invasive intrusions of minor or massive scale. The U.S. government has launched an intensive effort to reduce the number of attacks and make computing more secure and safe, but it must get continually smarter and faster to stay ahead of the cybersecurity threats. Interoperable data transactions can be a boon to this effort in terms of wide-area alerting, information exchange between the parties working on the problems, and as a means to guarantee that the correct distribution was made to fix the problem.

4. **Trust.** Trust is a very complex element of our online world (as well as our everyday, regular life), and is a component of many of the other issues we've examined. Knowing who you are dealing with, what the rules are in the short term, where the data is going in the long term, and other factors all contribute to trust. It is far easier to trust someone with whom you've had

Chapter Nine

a personal relationship than it is to go to an unknown site and hope for the best. Would you rather buy a sophisticated piece of equipment that might require a return through Amazon. com, or through an unknown lowest bidder? On an infallible commodity product costing $39, it might make no difference, and so the required trust level can go down. Trust is by definition a multi-party interaction, so interoperable data can help structure the information exchange between the parties, and enable far better and more standardized audit trails for analyzing trust violations.

5. **Information overload.** Human factors will remain a critical limitation. Humans aren't getting much smarter or more capable, but our collective intelligence is producing more data every day. Finding the right stuff in the avalanche of information is hard, and will be a persistently growing, big problem. Interoperable data can assist in this area long term by building structured filters that look at incoming information and make fine-grained decisions about how it should be handled. In addition, more interoperable transactions can save large amounts of time (a simple example is streamlining calendar updates), giving people more time to work on higher-priority issues.

6. **Device synchronization and information access.** My email and calendars are well synchronized, but the rest of my applications and devices are a hopeless mess, and only my memory can tie it all together. Changing devices can be a nightmare if you don't have the skills to move data and ensure its accuracy and completeness. Much work needs to be done here to allow us to be organized, and to allow us control

over who has what degree of access to our information – family, friends, colleagues or strangers – under a variety of circumstances. Interoperable data will help streamline this information, as well as enable more sophisticated exchange between individuals and better audit trails of the information on the different devices.

Some companies and products to watch

Here are a dozen public and private companies I'm watching because I think they are innovative and smart, and have fantastic technology and opportunities for the future.

1. **American Systems** is one of the larger employee-owned companies in the U.S. It is a system integrator with a wide range of capabilities and services, serving the U.S. government market.

2. **Anakam** is an emerging player in the authentication arena. It has developed very innovative two-factor authentication technology that's been well received and is very easy to use in securing web portals.

3. **Apple** is an enigma – it makes such noteworthy products that people are willing to lock themselves into proprietary environments to use them. I'm impressed with Apple's design ability and it has raving fans for almost all of its products. Apple is always exciting to watch in all the markets it participates in.

4. **Cisco** is growing in importance as networks become more interconnected and need performance, security, collaboration, and more at the network level. The company is a perennial performer, with advanced technologies that will be critical to the future of the networks. Their advanced videoconferencing application, Telepresence, is amazing.

5. **The Collaborative Software Initiative.** Open source software is a burgeoning area, and CSI is working to develop a model that will allow free software adoption with solid upgrade and support capability. One of CSI's initiatives is TriSano, a community-based, citizen-focused surveillance and outbreak management system for infectious disease, environmental hazards, and bioterrorism attacks.

6. **CommsFirst** is a security and emergency management company that specializes in deploying people equipped with communications technology into the field, using mobile networks, sensors, and situational awareness capabilities in communications-equipped Hummers or portable communications stations.

7. **Gold Class Cinemas,** part of the Australian film producer and entertainment company Village Roadshow, combines film with first-class accommodations (reclining chairs, pillows, blankets, waiter-summoning buttons) and first-class food and ambiance to create a unique movie experience. On the business side, creating movies has become increasingly data-driven, and since movies moved to digital formats, so has distribution. Gold Class is realizing efficiencies and new capabilities with interoperable formats, and developing new ways to use interoperable data to strengthen consumer loyalty and sell-through.

8. **Google** has become a company easy to admire and follow – it's had strong growth, impressive press, and waves of innovative technology without the monopoly problems (so far) that Microsoft had to deal with as it matured. They are changing the computing world in a number of different directions, and will be a player in the interoperable data world as it matures.

9. **PointAbout** has developed an application framework for mobile phones, particularly the iPhone, which allows applications to be produced and deployed very rapidly for the customers they service.

10. **SAIC** is a $12 billion government and commercial system integrator that develops, combines, and deploys technologies worldwide. It's doing cutting-edge work on the Smart Grid, health care records, and much more.

11. **Siemens** has been beaten up in recent years with corruption scandals and associated SEC issues, but it has a broad, sweeping vision around an emerging concept called "megacities" – the idea that population will continue to concentrate in cities, leading to ever more cities of over 10 million (18 qualified in 2000; 25 in 2009). To operate efficiently, a city of this size must have many smart subsystems (transportation, health, water, energy, and more) that all communicate with one another.

12. **Signacert** has built and refined technology that will help in the cybersecurity battles. Known as "whitelisting," their technology ensures a stable software environment on devices by creating a baseline that can be compared to the current environment at any point.

Chapter Nine

Chapter Nine summary

1. The future will be here before we know it – so planning and thinking about it *now* is critical for you, your organizations, and the world. Think interoperable data.

2. Our future is becoming complex and intertwined, with many issues interacting in ways we wouldn't have anticipated. Think of interoperable data as a means to overcome the challenges.

3. Standards can allow very rapid change – a new web site can burst on the scene in days because the plumbing has already been created and tested by someone else, and stands ready for repurposing.

4. Follow the cost curve by looking at what is unique and expensive today – and imagine it being free or near-free in the future, no matter how absurd that may sound.

5. Interact and share with other people, and form a collective that pays attention to the trends you care about. Having many eyes watching the future will reduce the time individuals spend and ensure a broader view.

6. Follow the futurists' and prognosticators' predictions and use them as a guide. Realize that most of them will fail most of the time, and that things always take longer than predicted. Some of the best futurists (from a practical standpoint) will be the 10- to 20-year olds – they'll be the early users of what the world will adopt.

Chapter Ten

Raves and conclusions

Be who you are and say what you feel, because those who mind don't matter and those who matter don't mind.
—Dr. Seuss

I'd like to acknowledge a few of the great people and organizations I've encountered along my interoperable data journey over the last few years. All of them are notable in one or many ways; this praise is for their efforts to drive the world forward and create systems that will support the next generations of humanity.

Kudos (in no particular order)

- **OASIS** (Organization for Advancement of Structured Information Standards). What a superlative organization. I've interacted with many of the people there, and they have all been effective, enthusiastic, and passionate about the work they are doing in the field of standards and structured information. OASIS has dedicated employees, diverse membership, and a willingness to make things happen as quickly as worldwide consensus-building allows. I have been involved in a few unsuccessful standardization efforts by other organizations, and the difference between OASIS and others is quite distinct.

- **NIEM** (National Information Exchange Model). The Department of Justice, in partnership with the Department of Homeland Security, has spawned the National Information Exchange Model structure. This is a very important initiative that will have far-reaching consequences across all government as well as commercial and international markets. The people working this have been real pioneers, toiling upwards toward success, and deserve a lot of credit for persistence. The agency leadership and Congress should increase the budgets dramatically for these standardization efforts, to produce the same kind of infinite return that the Internet efforts in DoD created.

- **NOAA.** The U.S. National Oceanic and Atmosphere Administration (colloquially, "the Weather Service") has been a pioneer, cutting-edge participant in sharing data with companies and the public. NOAA was an early adopter of the Common Alerting Protocol, and was invaluable in my company's efforts to build a common operating picture that merged weather with many different threats.

- **President Obama, Vivek Kundra, and their transparent government initiatives.** President Obama deserves a nod of thanks for making the idea of a transparent government mainstream, but every major idea needs an implementer. As the first National Chief Information Officer, Vivek Kundra is moving the government toward a more open way of doing business, which allows organizations to receive access to raw data. This can be used to sanity-check and analyze government programs and spending. For example, there is a strong commitment to show the expenditures of the American Recovery and Reinvestment Act of 2009 (the economic

Chapter Ten

stimulus package), from beginning to end, starting with the government allocation of dollars, distributions to the states and major government agencies, municipalities, contractors, universities, and subcontractors. (See www.recovery.gov.) The goal is to see how the money was used to drive jobs and new economic initiatives at the grass-roots level. As (and if) this mentality takes hold, much of government can be opened up to the tax-paying public, allowing feedback and better use of our government dollars. This effort is still a work in progress, but I am confident that their efforts will succeed.

- **The creators of the Common Alerting Protocol.** The Partnership for Public Warning, OASIS, and the Emergency Interoperability Consortium did the initial work on the Common Alerting Protocol, and the number of lives saved over the coming years will be a direct result of their efforts. They should be very proud of their efforts.

- **World Meteorological Organization** (WMO). The WMO has picked up the CAP standard and is driving its adoption across the world of weather outside the United States. Eliot Christian, previously with the U.S. Geological Survey, has done several coordination meetings, and is also driving a registry of global weather information providers – a much needed, centralized authority where information is going to be distributed to and from many sources.

- **Global Humanitarian Foundation – Weather Info for All.** Kofi Annan, former head of the United Nations, announced this initiative in 2009 at a conference in Geneva. This effort, though simple in scope, could save many lives and prevent property damage in economies of all types.

Silver Bullets

- **The IJIS Institute,** headed by Paul Wormeli, is a groundbreaking public/private partnership effort. They have championed NIEM, driven several standards to life, trained a host of people on interoperable data structures and XML, and worked to support both industry and government. IJIS is funded primarily through grants from the U.S. Department of Justice, the Bureau of Justice Assistance, and the U.S. Department of Homeland Security.

- **Sunlight Foundation.** The Sunlight Foundation is a non-profit in Washington, D.C. that really stands out. For the first half of 2009, they seemed to be everywhere, doing much good. They drove Transparency Camp and brought together hundreds of people interested in opening up government data to the public and organizations. They also have a strong, open source-oriented development effort that has contributed to many initiatives.

- **Bar Camp format.** I attended several exciting "non-conferences" in the spring of 2009. Transparency Camp, Government 2.0 Camp, and Crisis Camp were all held using a unique format – free or nominal attendance fees, and an agenda that was created in real time by the participants. These forums attracted some of the best and brightest people I'd seen at conferences. The lack of structure allowed flow and the ability to get rapid exchange around emerging topics.

- **Golden Phoenix** 2008 leadership and participants. I have participated in a number of public safety/homeland security events, exercises, and demonstrations conducted by various agencies, but the commitment, spirit, and free-form approach of this event set it apart as an excellent effort.

Chapter Ten

- **Google** gets raves on a number of fronts.
  - Releasing KML to the Open Geospatial Consortium – this is a gift to humanity that will pay a lot of dividends over the coming years. Geospatial information has been the domain of the experts for a long time, but it's moving out to the masses through mashups, easy-to-use interfaces, and increasingly standardized data.
  - "20 percent time" projects for employees: Google gives its people discretionary time to drive efforts they are passionate about.
  - The Google foundation: Google provides substantial funding to accomplish good things through their foundation.
  - Fostering competition: Everyone in the industry is moving faster because of Google's efforts on so many fronts.
- **Disaster Risk Reduction** (DRR) committee. I attended a DRR day hosted by the Red Cross and attended by most of the major NGOs (non-governmental organizations) and non-profits that have a humanitarian assistance mission. By focusing on the preparation for and prevention of disasters, they are helping many developing economies take the steps necessary to plan for emergencies before they strike – a vastly superior method.
- **USGS** (United States Geological Survey). Like NOAA, this group has been one of the leaders in collecting and dispensing information. Their earthquake feed in Common Alerting Protocol is freely available in real time.

- **NIUSR.** The National Institute for Urban Search and Rescue, led by Lois McCoy, has been working the emergency communications and readiness issues for a long time. They deserve a huge set of kudos for their perseverance and commitment.

- **www.consortiuminfo.org.** Sponsored by legal firm Gesmer Updegrove LLP, this highly useful, bimonthly publication is focused on standards and open source software. I'd recommend subscribing if you've become interested in the power of standards. Andrew Updegrove contributes compelling articles on standards and the thought processes surrounding the use of standards.

Conclusions and observations

The following is a collection of somewhat random thoughts that occurred to me while writing this book. They are opinions, not facts, and all mine (though friends and reviewers have suggested things I've incorporated). These are not in any order of importance or impact. A few items sound dire, but I have faith in humanity and believe that the world will overcome its current challenges to emerge as more capable and more democratic.

- **The world "change rate" is rising.** We're living in a time of increasing volatility; you are going to see significant changes over the coming 10 years, and unimaginable change over the next 50. We're moving so fast that some of these changes could be very bad, but the pace isn't going to slow down anytime soon. The optimist in me bets that there will be discoveries that will take care of today's big problems, but that more difficulties will emerge. Bette Davis said it best: "Fasten your seat belts. It's going to be a bumpy night."

Chapter Ten

- **More smart people are alive than ever before.** And because of the instant reach of technology around the world, we're harnessing the collective intelligence of the technology and technology-empowered communities in ways that have never happened in history. Almost any presentation can be posted on the web at www.slideshare.net, videos of every conceivable type show up on YouTube, new documents show up in search engines, and structured and linked data will continue to climb the ladder of importance. Thanks to medical advances and a focus on exercise and diet, many smart people are living longer and making more contributions to the world. In America, baby boomers still have the opportunity to make huge contributions toward humanity, knowledge and the world at large – and as one of those boomers, I'm challenging everyone to do this in addition to (or instead of) enjoying retirement.

- **English language dominance.** What country has the largest number of English-speaking citizens? "China" is the right answer. Having one language adopted and used globally will have positive impacts on the world at large. Translation capabilities, such as Google Translate, will help bridge those gaps.

- **Interoperable data can help define the next generation of applications** that will enable sharing of information. While we've made tremendous progress over the last 25 years, much of our information is still very disjointed, unstructured, and unusable, requiring a significant amount of work. As simple interoperable data formats take hold, followed by linked data and a full implementation of the Semantic Web, the pace of change in the data will be key.

- **The interoperable data wave will create winners and losers.** Like most major changes, there will be companies that adapt to the new ways of doing business…and those that cling to the old ways. Some will not change because their software is too brittle to be redeveloped. Others will fight to retain proprietary advantages (versus being open and adaptable). One immense contest of wills will come in the smartphone market between Apple and Google. Apple's iPhone set the bar for ease of use, applications, and a host of other elements, many enabled by its excellent portable computing and communications platform. So what's not to like? Consider: The iPhone (and the iPod with iTunes) is a totally proprietary solution, with Apple as the sole dictator of policy, technology, and governance. Google, on the opposite end of the spectrum, has released the Android phone, which is totally open to outside developers, able to be modified, and encourages free-form innovation by anyone. It will be a very high-profile and hard-fought contest, but I'm betting on Google or a Google partner to ultimately win the day over Apple's proprietary capabilities. Check back with me in 2015 for an update. It is an incredibly important race, and the competition will make everyone better off.

- **We'll move toward solving information overload.** The amount of information is rapidly expanding, and we humans are not scaling our native brainpower to handle it. However, I'm confident that through interoperable data, advances in software that provides targeting and filtering, real-time data exchanges driven off your mobile device, and other innovations, we will move toward a very context-sensitive environment in which you are far more aware of the things you want to know, and far more able to screen out the vast amounts of unimportant information that bombard you today. It won't be perfect, but it will be better.

One critical assist will come from the long-planned Semantic Web. As the web is structured now, humans can read its information and process it, but machines cannot. The Semantic Web seeks to (among other things) develop languages for expressing information in a machine processable form, so that the machine processes that drive search and other web uses will become smarter and deliver better, more filtered, results. The bad news is that making the Semantic Web a reality will require a lot of re-learning by developers, and re-engineering the legacy software will cost billions of dollars. The good news is, like the shift from the telegraph to the telephone, the entire world will change for the better. The book *Pull: The Power of the Semantic Web to Transform Your Business*, by David Siegel, illustrates some of the great benefits that can come from semantic thinking.

- **Population changes will force smarter thinking** and more collaboration. In all the industrialized nations, birthrates are plummeting, native populations (non-immigration) are projected to level off or decline sharply over the next 50 years, and the remaining populations will get older on average. At the same time, exurban populations continue to migrate to the cities, making metropolises ever larger and denser, and more in need of well-coordinated services. These changes will cause more automation, more electronic decision making, and more communications capabilities. Humans are at home with unstructured data and concepts; robots and other time-saving devices will need more structured directions.

- **Human nature stays pretty constant.** It is easy to get caught up in all the peripheral changes that have occurred and continue to develop, and somehow assume that people will change along with it. I don't think this is the case – we're still basically the same as people 500 years ago, 1,000 years ago, or 10,000 years ago. Unless someone comes up with new pharmaceuticals or genetic modifications that somehow reshape humanity, we'll still be dealing with the whole collection of personalities, eccentricities, and variations that we do today. Maybe if the playing field levels off, we'll reduce a bit of the hopelessness that seems to drive many of the conflicts today, but I'm not optimistic.

- **Health care won't get better without good standards.** I've watched the numbers measuring the waste and overhead of our current health care system. The policy issues are huge, but no costs will be cut in health care without good, widely adopted standards that span all the providers, insurance, individuals, and governments involved in this essential service. If the policy doesn't get fixed, all the data standards in the world won't help. However, good policy working with poor underlying data will doom the industry to another decade of massive cost overruns.

- **Energy is another key battlefield.** The current U.S. electrical grid is a mess and getting more unreliable as time progresses. Rebuilding the physical infrastructure is critically important, but once again, having the ability to standardize, share, and leverage data about the issues inside the current grid and the emerging Smart Grid is essential. The good news is that the current grid can be enhanced while the new technologies are being conceived, tested, and deployed – enabling major savings, increased utilization, better reporting of blackouts, and enhanced visibility by all the parties that need information to keep things running while we figure out a long-term strategy and implement it across the world. Let's keep the lights on for the next generation.

- **Climate change will force major innovation.** Along with energy, global warming will force the world's nations to come together to achieve a planetary solution. It may take too long and not be enough (if you are reading this book underground in a bunker or on another planet – it clearly took too long). Barring that low-probability scenario, the world will need to find a way to counteract the changes that seem to be accelerating in terms of glaciers melting, severe weather increasing, and temperatures rising. A key element in this will be the ability to collect and share data in real time that can also be added to climate models that both look back and project forward months, years, decades, and millennia.

- *The last comment.* I hope you enjoy reading this book as much as I enjoyed writing it. Best wishes to you for an interoperable future.

—The End—

Appendix A

Chapter notes and suggested reading

Introduction

Sources

Brooks, Frederick. *The Mythical Man-Month: Essays on Software Engineering.* Reading: Addison-Wesley Professional, 1975.

Chapter One

Sources

Commission on the Prevention of Weapons of Mass Destruction Proliferation and Terrorism. "The Clock is Ticking." URL http://www.preventwmd.gov/static/docs/report/WMDRpt10-20Final.pdf (February 6, 2010)

Congressional Report H. Rpt. 109-377. "A Failure of Initiative – Final Report by the Select Bipartisan Committee February 15, 2006." URL http://www.gpoaccess.gov/serialset/creports/katrina.html (February 6, 2010)

Chief Information Officer, Department Of the Navy. Memorandum 11 October 2002: "Interim Policy on the use of XML – 2002 (http://xml.coverpages.org/DON-XML-Memo2002.pdf)

Stephenson, W. David. *Democratizing Data* (release date Summer 2010)

Chapter One, continued

Companies and organizations noted

Facebook (www.facebook.com)

GAO (www.gao.gov)

Google (www.google.com)

LinkedIn (www.linkedin.com)

Magellan (www.magellangps.com)

National Oceanic and Atmospheric Administration (NOAA) (www.noaa.gov)

Skype (www.skype.com)

Twitter (http://twitter.com/)

Yahoo (www.yahoo.com)

U.S. Department of Defense (www.defense.gov)

U.S. Department of Energy (www.energy.gov)

U.S. Department of Homeland Security (www.dhs.gov)

U.S. Securities and Exchange Commission (www.sec.gov)

Chapter Two

Sources

Consortiuminfo.org. "Standards Bulletin #29 – Thinking about standards inside the Box , May 2006." URL http://www.consortiuminfo.org/bulletins/may06.php#considerthis (February 7, 2010)

Appendix A

Chapter Two, continued

Alder, Ken. *The Measure of All Things: The Seven-Year Odyssey and Hidden Error That Transformed the World.* New York: Free Press, 2002.

Companies and organizations

McDonald's (www.mcdonalds.com)

Walmart (www.walmart.com)

Chapter Three

Sources

O'Dell, Peter. *The Computer Networking Book.* Chapel Hill: Ventana, 1989.

Companies and organizations

Apple (www.apple.com)

Compaq (www.compaq.com)

Digital Equipment Corporation (DEC) (www.digital.com)

Google (www.google.com)

Hewlett Packard (www.hp.com)

IBM (www.ibm.com)

Intel Corporation (www.intel.com)

Lotus (www.ibm.com/software/lotus)

Microsoft (www.microsoft.com)

Chapter Three, continued

Napster (www.napster.com)

National Center for Supercomputing Applications (NCSA) (www.ncsa.illinois.edu)

Netscape (www.netscape.com)

Radio Shack (www.radioshack.com)

VisiCalc (www.bricklin.com/visicalc.htm)

Wired magazine (www.wired.com)

Xerox (www.xerox.com)

Chapter Four

Sources

Hock, Dee. *One from Many: VISA and the Rise of Chaordic Organization.* San Francisco: Berrett-Koehler, 2005.

Wikipedia. "EZ-Pass history and implementation." URL http://en.wikipedia.org/wiki/E-ZPass (February 7, 2010).

Companies and organizations

American Airlines (www.aa.com)

Caterpillar (www.cat.com) E-Z Pass (www.ezpass.com)

Digital Equipment VAX (www.digital.com)

GM (www.gm.com)

IBM (www.ibm.com)

Appendix A

Chapter Four, continued

Ingram Micro (www.ingrammicro.com)

Quicken (quicken.intuit.com)

VISA (www.visa.com)

Chapter Five
Sources

OASIS.org (http://www.oasis-open.org/home/index.php)

W3c.org (http://www.w3.org/XML/)

Wikipedia (http://en.wikipedia.org/wiki/XML)

Companies and organizations

Georgetown University (www.georgetown.edu)

MyCareTeam (MCT) (www.mycareteam.com)

National Library of Medicine (www.nlm.nih.gov)

U.S. Army Medical Research and Materiel Command (www.usamraa.army.mil)

Chapter Six
Sources

Von Oech, Roger. *A Kick in the Seat of the Pants.* New York: Harper Perennial, 1986.

OASIS. "CAP 1.1 specification." URL http://www.oasis-open.org/committees/download.php/.../emergency-CAPv1.1.pdf (February 7, 2010)

Silver Bullets

Chapter Six, continued

Open Geospatial Consortium. "KML 2.2 specification." URL http://www.opengeospatial.org/resource/products/byspec/?specid=284 (February 7, 2010)

Companies and organizations

CNN (www.cnn.com)

Digital Equipment Corporation (DEC) (www.digital.com)

ESRI (www.esri.com)

International Telecommunications Union (www.itu.int)

National Science and Technology Council (www.ostp.gov/cs/nstc)

Partnership for Public Warning (www.partnershipforpublicwarning.org)

RSOE/Havaria (hisz.rsoe.hu)

Silicon Graphics (www.sgi.com)

Swan Island Networks, Inc. (www.swanisland.net)

Tsunami warning center: West coast and Alaska, NOAA tsunami warning center,(http://wcatwc.arh.noaa.gov)

Tsunami warning center, Hawaii. NOAA tsunami warning center (www.prh.noaa.gov/ptwc)

U.S. Department of Defense (www.defense.gov)

U.S. Geological Survey (USGS) (www.usgs.gov)

ViaLogy (www.vialogy.com)

Warning Systems Incorporated (www.warningsystems.com)

WEB-EOC (www.esi911.com)

Appendix A

Companies and organizations

Association of Public Safety Communications Officials (APCO International) (www.apcointl.org)

Central Station Alarm Association (CSAA) (www.csaaul.org)

Facebook (www.facebook.com)

Global Humanitarian Foundation (www.humanitarianorg.com)

Health Level Seven (HL7) (Health Level Seven)

LinkedIn (www.linkedin.com)

National Institute for Standards and Technology (NIST) (www.nist.gov)

Oak Ridge National Laboratory (ORNL) (www.esd.ornl.gov)

SAIC (www.saic.com)

Sensorpedia (www.sensorpedia.com)

Slideshare.net (www.slideshare.net)

Squidoo (www.squidoo.com)

Stimulus.org (http://stimulus.org/)

Twitter (http://twitter.com/)

United Nations/Centre for Trade Facilitation and Electronic Business (UN/CEFACT) (www.unece.org/cefact)

U.S. Department of Homeland Security (www.dhs.gov)

Wikipedia (www.wikipedia.org)

World Meteorological Organization (WMO) (www.wmo.int)

Unified Incident Command and Decision Support (UICDS) (www.uicds.us)

Chapter Eight

Companies and organizations

Adobe (www.adobe.com)

Google (www.google.com)

Salesforce (www.Salesforce.com)

Chapter Nine

Sources

Gates, Bill. *Business @ the Speed of Thought: Succeeding in the Digital Economy.* New York: Business Plus, 2000.

Companies and organizations

Amazon.com (www.amazon.com)

American Systems (www.americansystems.com)

Anakam (www.anakam.com)

Apple (www.apple.com)

Cisco (www.cisco.com)

Collaborative Software Initiative (www.csinitiative.com)

CommsFirst (www.commsfirst.com)

Google (www.google.com)

IBM (www.ibm.com)

Oracle (www.oracle.com)

PointAbout (www.pointabout.com)

Chapter Nine, continued SAIC (www.saic.com)

SAP (www.sap.com)

Siemens (w1.siemens.com)

Signacert (www.signacert.com)

TRUSTe (www.truste.com)

Chapter Ten

Companies and organizations

Apple (www.apple.com)

Google (www.google.com)

IJIS Institute (www.ijis.org)

National Institute for Urban Search and Rescue (www.niusr.org)

NOAA (www.noaa.gov)

Open Geospatial Consortium (www.opengeospatial.org)

Sunlight Foundation (sunlightfoundation.com)

U.S. Bureau of Justice Assistance (www.ojp.usdoj.gov/BJA)

U.S. Department of Justice (www.justice.gov)

U.S. Department of Homeland Security (www.dhs.gov)

U.S. Geological Survey (USGS) (www.usgs.gov)

World Meteorological Organization (WMO)

YouTube (www.youtube.com)

General references

Specifications

CAP 1.1 specification, OASIS (http://www.oasis-open.org/committees/download.php/.../emergency-CAPv1.1.pdf)

bEDXL/DE, OASIS (http://docs.oasis-open.org/emergency/edxl-de/v1.0/EDXL-DE_Spec_v1.0.pdf)

EDXL/RM (http://docs.oasis-open.org/emergency/edxl-rm/v1.0/os/EDXL-RM-v1.0-OS.pdf)

EDXL/HAVE (http://docs.oasis-open.org/emergency/edxl-have/os/emergency_edxl_have-1.0-spec-os.pdf)

Standards

Alder, Ken. *The Measure of All Things: The Seven-Year Odyssey and Hidden Error That Transformed the World.* New York: Free Press, 2002.

ConsortiumInfo.org (www.consortiuminfo.org)

Semantic Web

DaConta, Michael, Leo J. Obrst and Kevin T. Smith. *The Semantic Web: A Guide to the Future of XML, Web Services, and Knowledge Management.* Hoboken: Wiley Publishing, 2003.

Siegel, David. *Pull: The Power of the Semantic Web to Transform Your Business.* New York: Penguin, 2009.

XML

OASIS.org (http://www.oasis-open.org/home/index.php)

W3C.org (http://www.w3.org/)

NIEM.gov (http://www.niem.gov/)

The Open Government Initiative (http://www.whitehouse.gov/Open)

Appendix A

Appendix B

Best of NIEM Awards 2009 Inaugural

For the first time ever, the NIEM Program Management Office (PMO) presented five Best of NIEM Awards at the 2009 NIEM National Training Event. The awards were announced by Donna Roy, Executive Director of the NIEM PMO, and presented by Kshemendra Paul, Chief Architect at the Office of Management and Budget (OMB). The awards went to NIEM implementation projects that demonstrate how intergovernmental collaboration and innovative technology deliver results that increase government transparency, improve performance, and enable civic engagement. All the projects have been operational since 2008 and have reported specific measurable results. The awardees were selected because they leverage best practices and deliver innovative solutions effectively.

Collectively, the 2009 winners serve or process more than 16 million transactions per year. They have complex environments with legacy systems and use innovative new technologies. They integrate data across hundreds of data sources. Each winner includes collaboration across at least five agencies or teams. They represent great strides in information sharing with NIEM.

The 2009 "Best of NIEM" honorees are:

USCIS Enterprise Service Bus (ESB) Program

The United States Citizen and Immigration Services (USCIS) is the government agency that oversees lawful immigration to the United States.

Silver Bullets

Accomplishment

U.S.CIS receives and processes 7.5 million applications and petitions per year for more than 50 types of immigration benefits. The current process of receiving and processing these applications and petitions is paper-intensive, making it difficult for U.S.CIS to efficiently process immigration benefits. These forms are managed by different case management systems within U.S.CIS based on form type. Getting the forms into the disparate, stovepiped systems is just as challenging as getting the information out. The U.S.CIS Office of Information Technology has leveraged and reused the NIEM schemas and data models provided by NIEM.gov. Some services have very complex data requirements and required modeling more than 2,000 elements per form. Across the 80 forms, the 80 percent overlap of element data from each form allowed for significant reuse of the NIEM IEPDs. The use of associations and references is a vital best practice leveraged by these NIEM message exchanges to help manage the complexity and interdependency of the domain data model.

HHS-Connect, Information Architecture and Development

CONNECT is a consortium of five states that have agreed to pool their collective expertise to make interstate information sharing a reality.

Accomplishment

New York City's Health and Human Service (HHS) agencies serve more than 2 million clients. Before the HHS-Connect program, case workers were required to log in to several agency systems to view the clients' cases across the diverse benefit programs. To alleviate this, HHS-Connect now uses ground-breaking and innovative technologies to improve the city's ability to serve its HHS clients, while providing better customer service and online access. HHS-Connect leverages the technology resources in place at the city's

Department of Information Technology and Telecommunications (DoITT), furthering the implementation of PlanIT—the city's IT strategy. The varied types of data involved in the information exchange and the number of agencies affected by this exchange pose an unprecedented challenge to the IT services of DoITT. With the establishment of the NIEM exchanges, the worker portal is able to retrieve relevant client information from the connected agencies and collate it for presentation to case workers. The worker portal presents a holistic view of the client information across agencies to the case workers, allowing them to practice collaborative case management and make speedier decisions for benefit delivery.

Disaster Assistance Improvement Program (DAIP) Program Management Office

The Disaster Assistance Improvement Program (DAIP) exists to ease the burden of victims by creating a single access point for more than 40 federally funded forms of assistance (FOA). DAIP will consolidate benefit information, application intake, and status information into a unified system.

Accomplishment

Each year, approximately 50 disasters categorized as "Presidentially Declared" result in injury and death, destroy homes and businesses, and disrupt the lives of hundreds of thousands of people across the nation. The DAIP was designed to ease the burden of disaster victims by consolidating federally funded forms of assistance information, application intake, and status information into a unified system. Applications for assistance from 17 U.S. government agencies, including the U.S. Department of Homeland Security/Federal Emergency Management Agency (DHS/FEMA), runs across almost 60 forms, which are now available through a single, online application using NIEM to automate the exchanges. This new portal, DisasterAssistance.gov, eases the burden on disaster survivors and increases their access to disaster relief by creating a continually updated information clearinghouse that provides information on the benefits most valuable to disaster survivors, such as housing, food, and employment aid, in both English and Spanish. DisasterAssistance.gov reduces the time needed to apply for aid and check the status of claims while decreasing redundancy in application forms and processes.

Colorado Integrated Criminal Justice Information System

The Colorado Integrated Criminal Justice Information System (CICJIS) is an integrated computer information system that links five state-level criminal justice agencies—law enforcement, prosecution, courts, adult corrections, and juvenile corrections—to create one virtual criminal justice information system.

Accomplishment

The CICJIS program facilitates the sharing of critical criminal justice data among five state-level agencies at key decision points in the criminal justice process. It created the first technical enterprise sharing architecture in the state and is driven by the business information needs and business process requirements of Colorado's state criminal justice agencies. The partner agencies are the Colorado Department of Public Safety; the Colorado Bureau of Investigation; the Colorado Judicial Branch; the Colorado Department of Corrections; the Colorado Department of Human Services, Division of Youth Corrections; and the Colorado District Attorneys Council. To date, CICJIS has developed 35 transfers and 63 queries and processes more than 6 million transactions per year. However, the current architecture has design limitations that limited data sharing to the five partner agencies. The architecture had performed well for more than ten years, but because of its closed nature and the lack of security and limited scope of sharing to five agencies, it needed improvement. CICJIS identified two transfers that could be moved to the CICJIS Service-Based Architecture (SOA) solution without affecting the current architecture. CICJIS moved forward criminal justice data sharing using the Justice Reference Architecture (JRA) and NIEM.

Emergency Operation Center—Interconnectivity (EOC-i)

Paragon Technology Group is a fast-growing 8(a), woman-owned, small, disadvantaged business (WOSB, SDB) company headquartered in Tysons Corner, Virginia. Paragon has been recognized as a top 8a firm in Virginia, a top small business in the United States, and one of the 50 fastest-growing companies in the Washington, DC, area.

Accomplishment

Effective response to large incidents requires real-time collaboration among multiple agencies and jurisdictions. Emergency Operation Centers (EOCs), activated during an incident, use many different systems to support operations and situational awareness. Most EOCs are neither interoperable nor interconnected, which makes it very difficult to coordinate resources and inform the decisionmakers. The EOC-interconnectivity (EOC-I) project defined a set of data exchanges for requesting and responding to incident and resource information enacted and acquired during the incident. The NIEM-conformant exchange and prototype system is based on emerging Internet technologies and designed to improve information sharing, situational awareness, and collaboration by regional EOCs during multijurisdictional emergencies to maximize the situational awareness for first responders. The EOC-I project was developed through interactions with state, regional, local, and tribal first responders in the Seattle and Cincinnati regions as well as in coordination with FEMA National Incident Management System (NIMS) multiple working groups.

Appendix C

World Meteorological Organization Members

Press release, 27 November 2009

The World Meteorological Organization (WMO) has announced the launching of the "Register of WMO Members Warning Authorities."

The introduction of the Register is an important step towards achieving a "single official voice for dissemination of warnings", which is one of the priority areas identified by WMO Member countries and territories. The Register has replaced the "list of Members' legal basis for issuing weather warnings" that was previously posted on the WMO PWS Website.

The Register contains: country names; organizational name of the alerting authority; geographic area wherein the organization performs its alerting; types of messages for which the organization has authority; and, Internet URLs where the alerting authority serves its forecasts and / or alert messages. The Register is publicly accessible to all for viewing only. Persons from WMO Member countries who are designated to update the Register will access a password-protected version of the Register. The Register Website includes a Google map tool for visualizing the geographic area associated with each register record.

National Meteorological and Hydrological Services (NMHSs) have no greater responsibility than to ensure the safety of life, the protection of property, and the well-being of their nations' citizens. Since about 80 % of all disasters involve the weather, in most countries, the

NMHS is the key national agency for issuing warnings. Sometimes this agency also has responsibility for volcanic and earthquake hazards. In some countries, floods and hydrological forecasting are handled by a separate agency, for example, a river basin management authority. When a serious hazard is imminent, to avoid confusion, it is important that there be a single authoritative voice to issue warnings, and it is important that users of warnings can quickly determine who has the responsibility in a particular jurisdiction. In response to this need, the "Register of WMO Members Warning Authorities" has been designed to clarify the sources that each WMO Member has authorized to issue particular warnings.

The Register contributes to ensuring effective service delivery in the provision of official warnings and information of high-impact weather and extremes of climate, to government authorities, and emergency communities in order to aid them in their mission to protect the lives, livelihoods and property of the public they serve.

Governmental policy makers, emergency management bodies, and the media are therefore encouraged to use this Register to help them in their daily operations.

About the Author

Pete O'Dell is a technology executive based in Washington, D.C. He's had a long career on the information highway, working for companies including Digital Equipment Corporation, Autodesk, Microwarehouse, Microsoft, and Online Interactive, and cofounding Swan Island Networks. Pete cnsults for companies including SAIC, CommsFirst, HCLTEM, and Village Roadshow Limited.

Pete's first book, *The Computer Networking Book*, was published in 1989 by Ventana Press. He has authored numerous articles, and speaks at conferences on technology and business topics.

He served honorably in the U.S. Army, and has been involved in emergency management, situational awareness, and secure information since 9/11. He has an undergraduate degree in business and an MBA.

In his free time, Pete is an avid fly fisherman, a reasonable runner, a voracious consumer of technology information, and a poor golfer.

He can be reached at peterlodell@gmail.com. Comments on actions you took based on reading this book would be highly appreciated. If you like the book, please post a review online somewhere.

Photo by Pam Angelus

Index

Symbols

7.62mm cartridge   42
9-1-1   95, 97, 110, 136, 141, 143
9-1-1 Public Safety Answering Points (PSAP)   141
9/11   8, 16, 17, 18, 95, 113

A

ACORD   83
Adams, John Quincy   40
ADML (Architecture Description Markup Language)   82
airline reservation system   63
Amazon.com   7, 157, 182
Amber Alert   20
American Airlines   63, 64, 178
American Recovery and Reinvestment Act of 2009   164
American Red Cross   113
American Systems   158
ammunition   42
Anakam   158
Andreessen, Marc   54
Android (mobile operating system)   152
Annan, Kofi   121, 165
APCO (Association of Public Safety Communications Officials)   142, 181
Apple   52, 57, 59, 152, 158, 170, 177, 182, 183
ARCNET   56
assurance   19, 22, 23, 78, 121, 197
ATIS (Alliance for Telecommunications Industry Solutions)   86
authentication   19, 23, 67, 78, 154, 155, 158
Autodesk   106
automated feeds   14

B

Bar Camp   117, 166
Bell, Alexander Graham   5
Berners-Lee, Tim   54, 91, 145
Big Brother   155

biometrics 155
Botterell, Art 95
break bulk 43
Bressler, George 109
Brooks, Frederick 2
browser 7, 54, 55, 59, 110
Buckley, Justin 138

C

CAP (Common Alerting Protocol) xi, 88, 91, 92, 94, 95, 96, 97, 98, 99, 100, 101, 102, 104, 107, 109, 110, 111, 112, 113, 122, 165, 179, 184
  adoption 96
CASRAI (Consortia Advancing Standards in Research Administration Information) 85
Caterpillar 67
Central Station Alarm Association (CSAA) 142, 181
Cisco 158, 182
climate change 173
cloud computing 153
COBOL 50, 51
CODASYL 51
Collaborative Software Initiative (CSI) 159
common operating picture 95, 109, 164
CommsFirst 47, 48, 111, 159, 182
Compaq 52, 54, 177
consortiuminfo.org 168
Crisis Camp 166
CUI (Controlled Unclassified Information) 21
customer relationship management 135
cybersecurity 156, 160

D

data collections 133
DEC (Digital Equipment Corporation) 53, 54, 177, 180
DEC Rainbow 53
Dell, Michael 52
DHS (Department of Homeland Security) 33, 109, 121, 124, 164, 189
diabetes management xi, 71

Index

discoverability 22
DNS (Domain Name Service) 55
DocBook 84
DoD (Department of Defense) xv, 23, 109, 111, 164
DoE (Department of Energy) 21
DoJ (Department of Justice) 121, 164, 166, 183
DRR (Disaster Risk Reduction) 167

E

early computers 49
earthquake 12, 16, 96, 98, 110, 147, 167, 194
EAS (Emergency Alerting System) 122
ebXML (Electronic Business using eXtensible Markup Language) 119, 120
EDI (Electronic Data interchange) 66, 67, 120
EDXL/DE (Emergency Data Exchange Language Distribution Element) 102
"Effective Disaster Warnings" 94
electrical grid 26, 151, 172, 199
Energy Independence and Security Act (EISA) 123
English language dominance 169
ESRI 106
Ethernet 56
eticket 64
Expedia 64
extensibility 76
External Alarm Interface Exchange Standard 141
E-Z Pass 68, 69, 178

F

Facebook 12, 123, 126, 127, 128, 176, 181
fail fast 135
FedEx 7
FEMA (Federal Emergency Management Agency) 101, 122, 123, 189, 191
field verification 87
FITSML (Flexible Image Transport System Markup Language) 79
Flow-Matic 51
Fort Hood shootings 12
FOUO (For Official Use Only) 21
FSLTIPP (federal, state, local, tribal, international, public, and private) 17

G

GAO (General Accounting Office)   20, 176
Gates, Bill   11, 52, 148
  Business @ the Speed of Thought   148
General Motors   67
Gesmer Updegrove LLP   168
Global Humanitarian Forum   121, 165, 181
Global Incident Map   111
global warming   147
GML (Geography Markup Language)   102, 104
Gold Class Cinemas   159
Golden Phoenix   xi, 109, 113, 166
Google   xv, 13, 55, 91, 103, 104, 105, 106, 117, 123, 134, 150, 152, 154, 159, 167, 169, 170, 176, 177, 182, 183, 193
Google Earth   103
Google Maps   103
Government 2.0 Camp   166
government transparency   115, 187
GPS   98, 151
GraphML   75
Great Wall of China   38

H

H1N1   12
HAVE (Hospital Availability Exchange)   102, 184
health care   i, 23, 24, 126, 160, 172
Health Level Seven (HL7)   82, 125, 181
Hewlett Packard   54, 56, 177
hierarchical data   88
Hobgood, Bill   141
Hock, Dee   65
Hollerith, Dr. Herman   62
Hopper, Rear Admiral Grace   51
HTML (HyperText Markup Language)   7, 55, 78, 103, 105
Hurricane Gustav   47
Hurricane Katrina   12, 99, 113

Index

I

IBM   49, 50, 51, 52, 53, 55, 56, 62, 63, 67, 153, 177, 178, 182
IBM Token Ring   56
identity   154
IJIS Institute   166
IML (Instrument Markup Language)   82, 83
information overload   157
information sharing   1, 3, 4, 17, 18, 31, 34, 62, 89, 97, 99, 107, 112, 113, 121, 125, 153, 187, 188, 191
Ingram Micro   67, 68, 179
Intel Corporation   53, 63, 177
Intergraph   106
International Business Machines. *See* IBM
interoperability   i, xii, 3, 6, 8, 9, 18, 41, 42, 43, 48, 67, 68, 72, 82, 87, 102, 106, 115, 123, 126, 131
interoperable data   i, ii, xi, 1, 3, 4, 5, 7, 8, 16, 17, 18, 19, 22, 23, 26, 28, 31, 33, 34, 35, 45, 49, 55, 59, 61, 64, 67, 71, 73, 74, 91, 92, 107, 112, 113, 116, 118, 121, 128, 132, 134, 135, 136, 141, 145, 147, 150, 151, 153, 157, 159, 161, 163, 166, 169, 170
IPAWS (Integrated Public Alerting and Warning System)   122
iPhone   50, 128, 152, 160, 170
iPod   57, 59, 170
IPV6 (Internet Protocol Version 6)   151
ITU (International Telecommunications Union)   95, 180
iTunes   57, 170

J

Jefferson Parish, Louisiana   47
Jones, Elysa   xv, 95, 101
JSON (JavaScript Object Notation)   75

K

Keyhole Corporation   103
Kmart   67
KML (Keyhole Markup Language)   xi, 91, 103, 104, 105, 106, 107, 167, 180
KMZ   104
Kundra, Vivek   116, 164

L

Lacey, Chip   68
LegalXML   83
LinkedIn   12, 126, 127, 128, 155, 176, 181
London Underground bombings   12
Lotus 1-2-3   53

M

Macintosh   52
Magellan   13, 176
MathML   75
MCC (Microelectronics and Computer Technology Consortium)   82
McCoy, Lois   168
McDonald's   44
mega-cities   146
metric system   39, 40
Microsoft   i, 52, 53, 54, 59, 111, 128, 148, 159, 177
Microsoft Virtual Earth   111
Microsoft Windows. *See also* Windows
microwave   62
Mosaic   54
mpXML   81
MS-DOS   52, 53
multi-factor authentication   155
MusicXML   75
MyCareTeam (MCT)   71, 179

N

Napster   57
NASA   79
NATO   42
Nazis   42
NCSA (National Center for Supercomputing Applications)   54, 178
Netscape   54, 178
NIEM (National Information Exchange Model)   xii, 25, 88, 102, 121, 123, 125, 142, 164, 166, 184, 187, 188, 189, 191
NIST (National Institute for Standards and Technology)   123, 181
NITF (News Industry Text Format)   84

Index

NIUSR (National Institute for Urban Search and Rescue)   168, 183
NOAA ((National Oceanic and Atmosphere Administration))   33, 96, 98, 110, 116, 164, 167, 176, 180, 183

O

OASIS (Organization for the Advancement of Structured Information Standards)   xv, 83, 85, 94, 95, 96, 101, 102, 119, 120, 121, 163, 165, 179, 184
Obama, President Barack   24, 25, 115, 164
oBIX (Open Building Information Xchange))   79, 80
OGC (Open Geospatial Consortium)   103, 104, 107, 167, 180, 183
open source software   154, 159
Oracle   154
Orbitz   64
OTA (OpenTravel Alliance)   86

P

pandemic   147
Persano, LTC John   109
PIDX (Petroleum Industry Data Exchange)   84
Pierce, John   73
PII (Personally Identifiable Information)   19, 20
pilots   xii, 131, 132, 134, 135, 138, 139
PointAbout   152
policy   19, 24, 175
PPW (Partnership for Public Warning)   91, 94, 165, 180
privacy   19, 20, 31, 155, 156, 203
proprietary technologies   37
PSAP (9-1-1 Public Safety Answering Points)   141, 142
PSLX (Planning and Scheduling on XML Language))   84
punched cards   62

Q

Quicken   66

R

radio   8, 16, 18, 47, 48, 98, 112
Radio Shack   52
railroads   41, 42, 43

RELAX NG   84
RETS (Real Estate Transaction Standard)   85
risk   i, 13, 19, 22, 23, 31, 58, 87, 93, 131, 132, 134, 135, 136, 146, 148
RIXML   81
RM (Resource Messaging)   102, 184
RosettaNet   68, 80
RSOE/Havaria   111
RSS (Really Simple Syndication)   14, 75, 95

S

SABRE   63
SAIC   124, 160, 181, 183
Salesforce   135, 153, 182
Salk, Jonas   4
SAP   154
scalability   32
Scalable Vector Graphics   75
SDO (Standards Development Organization)   82
Sears   67
SEC (Securities and Exchange Commission)   26, 81, 125, 160
security   xv, 8, 20, 23, 65, 67, 80, 87, 95, 120, 124, 133, 135, 141, 146, 151, 153, 154, 155, 156, 158, 159, 166, 190, 204
Semantic Web   8, 169, 171, 184
sensitive information   20, 21, 204
Sensorpedia   123, 181
sensors   14, 89, 111, 150
Senusion   111
S-Expressions   75
SGML (Standard Generalized Markup Language)   75, 77, 84
shipping   32, 43, 45, 126
Siegel, David   171
Siemens   160
SIF (Schools Interoperability Framework [U.S.], Systems Interoperability Framework [U.K.])   80
Signacert   160
Skype   12, 176
Slideshare.net   128, 181
Smart Grid   27, 120, 123, 124, 149, 152, 153, 160
SNA (System Network Architecture)   55

Index

Social networking 126
source verification 87
Spencer, Percy 62
Squidoo 123, 181
standard gauge tracks 41
standards   i, ii, xi, xiii, 34, 35, 37, 38, 39, 40, 41, 44, 45, 47, 53, 54, 55, 56, 57, 59, 61, 64, 68, 75, 78, 79, 80, 81, 82, 83, 84, 86, 89, 91, 94, 102, 104, 105, 106, 107, 119, 120, 121, 123, 124, 125, 126, 129, 143, 145, 148, 163, 166, 168, 172, 176
status (of data) 21
Stephenson, David W. 27
"Strawberry Shortcake for 400,000" 93
structured data 28
success metrics 138
Sunlight Foundation 117, 166, 183
Swan Island Networks 8, 180

T

TCP/IP 55, 56
TDCC (Transportation Data Coordinating Committee) 66
technology trends 150
Telepresence 158
terrorism 149, 205
TIES 8, 9, 110, 111
timeliness (of data) 20
Transparency Camp 117, 166
trust   vii, 28, 65, 131, 156, 157, 205
TRUSTe 155
tsunami   ii, 16, 87, 92, 96, 98, 113, 147, 180
Tsunami Warning Centers 110
Twitter 12, 126, 127, 176, 181

U

UCC (Uniform Commercial Code) 44
UICDS (Unified Incident Command and Decision Support) 124
United Nations Centre for Trade Facilitation and Electronic Business 119
universal charger 58
Updegrove, Andrew 168

USB   58, 72
USGS (United States Geological Survey)   167
   Earthquake Notification Service   110, 206

V

VAN (Value Added Network)   67
verification (of data)   87
ViaLogy   111
VISA   61, 64, 65, 69, 178, 179
VisiCalc   53
Von Oech, Roger   93

W

W3C (World Wide Web Consortium)   75, 76, 77, 78, 84, 87, 184
Walmart   44
Warning Systems   101
Web 2.0   1, 8, 123
WEB-EOC   112
Weights and measures   39
WFS (Web Feature Service)   104
Whitney, Eli   40
WIFA (Weather Info for All)   121, 122
Wikipedia   123
Windows   52, 53, 55
Wired Magazine   xiii, 95
WMO (World Meteorological Organization)   102, 122, 165, 181, 183, 193, 194
WMS (Web Map Service)   104
Wolf, Gary   95
World Wide Web   7, 54, 55, 59, 75, 76
Wormeli, Paul   166

X

XBRL (eXtensible Business Reporting Language)   26, 81, 125
Xerox   52
XHTML   75

XML (Extensible Markup Language)   xi, 31, 61, 68, 73, 74, 75, 76, 77, 78, 79, 80, 81, 82, 84, 87, 88, 89, 91, 97, 101, 103, 105, 116, 118, 119, 120, 121, 125, 129, 141, 166, 175, 179, 184
  Key benefits of   76

Y

Yahoo   ii, 13, 176
YAML   75
yard, the standard for   39
YouTube   169, 183